Teacher Quality in Upper Secondary Science Education

Teacher Quality in Upper Secondary Science Education

International Perspectives

Edited by

Barend Vlaardingerbroek and Neil Taylor

First published 2016 by
PALGRAVE MACMILLAN

The authors have asserted their rights to be identified as the authors of this work in accordance with the Copyright, Designs and Patents Act 1988.

Palgrave Macmillan in the UK is an imprint of Macmillan Publishers Limited, registered in England, company number 785998, of Houndmills, Basingstoke, Hampshire, RG21 6XS.

Palgrave Macmillan in the US is a division of Nature America, Inc., One New York Plaza, Suite 4500, New York, NY 10004-1562. Palgrave Macmillan is the global academic imprint of the above companies and has companies and representatives throughout the world.

Hardback ISBN: 978-1-137-49088-9

Distribution in the UK, Europe and the rest of the world is by Palgrave Macmillan®, a division of Macmillan Publishers Limited, registered in England, company number 785998, of Houndmills, Basingstoke, Hampshire RG21 6XS.

Library of Congress Cataloging-in-Publication Data

Names: Vlaardingerbroek, Barend, editor. | Taylor, Neil, 1957- editor.
Title: Teacher quality in upper secondary science education : international perspectives / edited by Barend Vlaardingerbroek and Neil Taylor.
Description: New York, NY : Palgrave Macmillan, 2016. | Includes index.
Identifiers: LCCN 2015034935 | ISBN 9781137490889 (hardback)
Subjects: LCSH: Science teachers—Rating of—Cross-cultural studies. | Science teachers—Training of—Cross-cultural studies. | Teacher effectiveness—Cross-cultural studies. | Science—Study and teaching (Secondary)—Cross-cultural studies. | BISAC: EDUCATION / Evaluation. | EDUCATION / Secondary. | EDUCATION / Comparative.
Classification: LCC Q181 .T35127 2016 | DDC 507.1/2—dc23 LC record available at http://lccn.loc.gov/2015034935

A catalogue record for the book is available from the British Library.

Contents

Foreword

John Oversby

Upper secondary science education has a special place in the whole curriculum and in science education across different age ranges. It aims to be a preparation for future and continuing studies for some and to be a concluding episode for others. In many countries, it is the start of an obvious differentiated program based on the potential demands of highly varied future careers or on providing an understanding of the cultural input of science. It should not surprise readers that different countries have adopted different approaches, but it is interesting that there is so much similarity in content and aims. This is particularly so given the wide disparity in resources and wealth in the countries represented in this book.

If I were to explain a major purpose for upper secondary science education to outsiders, I might choose a building construction model to help me. A plasterer has been recruited to make the walls and ceilings of houses ready for the next stage of building. The plasterer is required, at one extreme, to work on houses that have been built by builders with limited skills and with recourse to only a limited pool of building resources. The walls of these houses are very uneven and not perpendicular to each other. The underlying material is patchy, with areas prone to collapse through water ingress. Some areas do not even attach to the rest of the wall. The plasterer endeavors to apply the plaster in the best way possible and, at first sight, the surface looks good enough to apply a coating. However, the slightest stress sees bits of the plaster coming away from the wall, making it very difficult for the plasterer to build the next layers. At the other extreme, other houses the plasterer works on have basic walls that are very well constructed and strong, thereby giving a highly coherent base for the plaster to stick on. The first coat looks good and ready to accept the next layer. It also goes on well, and so the process continues.

Of course, the skills of the plasterer make a considerable difference to the plastered state of the walls. High-quality plasterers—those who have knowledge of wall construction (good and bad) and wall surfaces, as well as sufficient experience and expertise—can make up for less satisfactory bases. This model might well explain many of the variations in the 18 country accounts of upper secondary science education so well described in this volume, and it connects well with the subject of teacher quality.

What makes for effective teaching of science education at this level of schooling? Naturally, we are highly unlikely to find or identify an approach that of itself can fully prepare students for the higher education phase of learning science. In a step toward finding a more comprehensive answer, the editors of this book begin with the core requirements that teachers of the sciences at the senior secondary level need to have:

- They need to be academically qualified in the science they teach.
- They need to have undergone thorough preservice training, including subject-specific pedagogy.
- They need to be operating in work environments conducive to achieving their professional goals.
- And they need to take part in meaningful continuing professional development (CPD) throughout their careers.

The evidence presented in the chapters of this book justifies the requirement for science teachers to be academically qualified in the science they teach, but it also shows that this requirement alone does not give us a clear idea of what constitutes that science or qualification. The chapter on Australia provides much research-based discussion of this issue, yet comes to a mixed conclusion about the value of a full science degree versus a degree that includes pedagogical study but a lower level of subject content knowledge. The Netherlands report seems to indicate a preference for an explicitly professional vocational track for science teachers, with teachers' content knowledge described in the traditional format of chemistry, physics, and biology. On the other hand, Yemen is reported as having a teaching force composed of predominantly unqualified people. The authors of the Yemen report also point to inequity in teacher supply, with more teachers in urban areas than in rural areas and considerably more male than female teachers. The authors furthermore suggest that international (external) input has been quite ineffective to date. There may therefore be more significant factors than academic qualifications at play here.

The reports from some of the other countries represented in this volume, such as Oman and Sri Lanka, provide only an overview of teachers' academic qualifications. The requirement to be academically qualified seems to be in need of further research concerning the priority of subject knowledge over pedagogical knowledge, the impact of the quality of subject knowledge, issues in teaching out of field, and concerns about recruitment. In contrast to data on content knowledge, data on pedagogical knowledge requirements for teachers are much more detailed and widespread in these reports. Nevertheless, it is difficult to see how effective these actions are in terms of improving teacher quality, and this is also the case in the data available from other international comparisons, such as the OECD's Teaching and Learning International Survey (TALIS) conducted on an iterative basis every few years (see, for example, the reports by Vieluf et al., 2012; Vincent-Lancrin et al., 2014). I echo the call for more critical scrutiny—through systematic and rigorous evaluations—of innovation.

The reports in this book additionally provide rich evidence of the work environments for teachers, the pressures under which they work, and, perhaps also of significance, the presence or lack of professional communities in which they can develop their professional competence. The latter is compounded, or supported, by programs of CPD that are directed toward not only supporting the achievement of national goals, such as improving quality, but also simultaneously engendering individual commitment by identifying and meeting personal goals for each teacher. Vincent-Lancrin and colleagues (2014) note that innovation is highly prevalent in education as a whole, from which we can assume that teachers are generally committed to change in pedagogy that is reasonable, manageable, and sustainable.

Given the declared purpose of upper secondary science education as a preparation for university in these reports, the similarity of content then becomes obvious. We should bear in mind that these are intended curricula and not necessarily enacted curricula in the classroom. There is clearly much more to learn about the role of teacher quality in making science education more effective.

The chapters in this volume bear testament to the rich evidence base that has been compiled by the authors. Taken together, the accounts provide a comparative narrative set within an overall structure of the state of upper secondary science education in each of the 18 jurisdictions. Central to the theme of the book, though, is the concept of teacher quality. This is a highly complex notion, linked to the concepts of content knowledge, methods of teaching, availability and deployment of resources, and more ill-defined ideas about classroom climate. In addition to being helpful from a comparative perspective, the chapters may also provide an imperative for seeking further information, for designing new research questions, as I have attempted to do here. A critical approach to reading each one in detail will bring many benefits.

References

Vieluf, Svenja, David Kaplan, Eckhard Klieme, and Sonja Bayer. 2012. *Teaching Practices and Pedagogical Innovations: Evidence from TALIS*. Paris: OECD Publishing, http://dx.doi.org/10.1787/9789264123540-en.

Vincent-Lancrin, Stéphan, Kiira Kärkäinen, Sebastian Pfotenhauer, Adele Atkinson, Gwénaël Jacotin, and Michele Rimini. 2014. *Measuring Innovation in Education: A New Perspective*. Paris: OECD Publishing, http://dx.doi.org/10.1787/978926421 5696-en.

Introduction: Upper Secondary Science Teacher Quality—"Those Who Can"

Barend Vlaardingerbroek and Neil Taylor

"Those who can, do; those who can't, teach."

—*George Bernard Shaw*

The great Irish iconoclast did not have a high opinion of the teaching profession. The schoolmaster of the nineteenth century was all too often the subject of mock respect rather than genuine esteem, and the Shavian view of the teacher as someone who could not succeed in his or her specialist field—the science teacher as someone not quite good enough to go into professional science, for instance—lingers on in the public psyche. But teaching *is* a profession (the International Labour Organisation classified secondary teaching as such, alongside doctors and lawyers, many years ago in its categorization of occupations), and akin to all professions, there are good practitioners and those who leave much to be desired. The focus thus shifts to those attributes that make for a good education professional, that is, teacher quality.

> Since teachers are the most valuable resource available to . . . schools . . . an investment in *teacher quality* and ongoing teacher professionalism is vital. In our view, this goal can only be realised by ensuring that teachers are equipped with subject-matter knowledge and an evidence- and standards-based repertoire of pedagogical skills . . . (Ingvarson and Rowe 2008, 6; original emphasis)

This book narrowly targets upper secondary science as the setting for an international comparative investigation into the complex, multifaceted area of teacher quality. Our preceding edited volume *Upper Secondary Science Education: Comparative Perspectives* (Vlaardingerbroek and Taylor 2014) elaborates on the premise that the upper secondary tier of schooling is a "special" one, for it is at this stage that the "broad, general education" paradigm gives way to specialized tracks aligning with post-secondary education and training programs, at least for some students. The book amply illustrates this transitionary role for science.

Upper secondary science in many countries represents a new academic elitism, wherein the "best" students enter science-intensive tracks at the upper secondary level, successful completion of which leads to admission to competitive-entry university programs, in turn leading to prestigious science-related careers. As a mechanism for the initial stages of the production of an indigenous scientific cadre, upper secondary science becomes a vital human capital development strategy for developing countries. The people entrusted with the task of grooming the young talent that will produce tomorrow's scientists, doctors, engineers, veterinarians, agronomists, and the like need to be far removed from the Shavian caricature.

For a start, they must be academically well qualified in their specialist fields. There is a convergence between the wealthy industrialized and poorer agrarian societies in this regard: whereas many primary teachers in the latter countries may have completed only lower secondary schooling themselves, upper secondary teachers are expected, like their colleagues in wealthier education systems, to be in possession of tertiary qualifications in their subject areas (UNESCO Institute for Statistics 2006). But the relationship between academic qualifications and teacher quality is, at best, nebulous. While there are certainly minimum thresholds—teachers cannot teach what they themselves have not mastered—there is no linear relationship between the two thereafter, at least not as measured by student learning outcomes (see, for example, Goe 2007; Harris and Sass 2011).

Various scholars and commentators (e.g., Eide, Goldhaber and Brewer 2004; Hattie 2003) describe teacher quality as the single most important school variable in terms of its influence on student achievement. In the words of Hattie (2003, 15): "Students who are taught by expert teachers exhibit an understanding of the concepts targeted in instruction that is more integrated, more coherent, and at a higher level of abstraction than the understanding achieved by other students."

And yet, as noted by Hanushek and Wößmann (2007, 65) in a World Bank paper, teacher quality is a difficult quantifier to pin down. Academic grounding, pedagogical knowledge, practical training, and continuing professional development all play a part, and the mutually supportive interactions among these make the contributions of individual components of the teacher quality equation elusive—if not pointless—to pursue in isolation.

While this consideration seems to be the case for well-resourced systems populated by qualified teachers, hair-splitting between, for instance, the effects on teacher quality of a Master's degree as opposed to a Bachelor's degree in a given subject is of little import in the context of resource-starved systems with many unqualified teachers. Barring poorly qualified and untrained teachers would leave teaching forces in many developing countries impoverished. On the other side of the ledger, as discussed by Ingersoll (2007), entry requirements to the profession may become a moot point when the job is regarded as an attractive one (in particular from a salaries point of view) and when standards rise as competition for admission increases.

What ultimately matters is how well students learn and how that learning is affected by the teacher quality multivariable. In broad terms, it seems a truism to posit that competent, confident teachers who know what they are talking about and who have been trained to do the job get better results than do subject-incompetent, insecure teachers who have been "thrown in at the deep end." Teacher education and training potentially make a lot of difference, but as Ingvarson and Rowe (2008) counsel us, we need to focus on teacher performance rather than on "paper qualifications"—for instance, not on what courses in pedagogy teachers have on their transcripts but how they go about teaching the subject. In the case of science, this focus requires us to ascertain whether a teacher is actually applying scientific inquiry in the classroom as opposed to paying lip service to a theoretical model he or she learned during training. Teacher professional behaviors accordingly become the functional interface between the theory imparted by training and classroom practice.

It is when we look at teacher quality through the lens of professional practice that we begin to see the potential of continuing professional development (CPD), still regarded by many skeptical practitioners as "time off" of little practical consequence (which it often is). As Whitehouse (2011, 10) observes, sustained, subject-specific, collaborative CPD programs with inputs from external expertise that are based on the learning needs of both teachers and students and that focus on classroom practice are effective because this combination leads to permanent changes in teacher practice (see also, for example, Everett, MacLeod, and Thurgood 2013 for the UK; Mokhele 2013 for South Africa). CPD has immense potential in the context of systems in which teachers are suboptimally qualified and trained, but it comes with high operational costs and needs to be "on the button" for that potential to be realized.

We regard it as axiomatic for the purposes of this book that upper secondary science teachers need to

- *Be academically qualified in the science they teach:* They should ideally be subject degree graduates at the level of specificity required. So, for example, a teacher of upper secondary chemistry should be a chemistry graduate. This proviso raises the issue of "out-of-field" teaching (e.g., a biology graduate teaching chemistry), a problem that is not confined to developing countries.
- *Have undergone thorough preservice training, including subject-specific pedagogy:* The traditional one-year program, such as the British PGCE, remains a benchmark, although there is a marked tendency toward better preservice training particularly in more affluent countries, including the double-degree model (concurrent Bachelor's degree in the subject area and in education) and the Professional Master's in education.

- *Operate in work environments conducive to achieving their professional goals:* These environments include not only the availability of material resources, such as school laboratories complete with the requisite equipment base, but also career structures that encourage and reward commitment and performance. This latter consideration invokes the existence of clearly defined performance standards, mechanisms for monitoring these standards, and mechanisms for following up performance evaluations. Professional associations may also be involved as part of the institutional support structure.
- *Actively participate in meaningful CPD:* This professional development needs to be ongoing throughout teachers' careers.

We make no distinction between a country with a per capita GDP of $50,000 and one with a per capita GDP of $500 with regard to this checklist. A medical or engineering graduate should meet certain universal competency criteria regardless of the World Bank rating of the country where he or she qualified, and we apply the same reasoning to the upper secondary education that precedes the university tracks leading to those specializations.

The full-fledged, modern upper secondary science teacher is competent in both science and in science education. He or she is a highly trained professional whose métier is the transmission of scientific knowledge and skills to emerging adults. *Those who can, do* teach upper secondary science, and they do it well, *providing* they have the means to do so. In the chapters that follow, we and our coauthors focus on those "means" as they apply to upper secondary science teachers in 18 countries spanning the global national income range, presented in descending order of per capita GDP.

References

Eide, Eric, Dan Goldhaber, and Dominic Brewer. 2004. "The Teacher Labour Market and Teacher Quality." *Oxford Review of Economic Policy* 20 (2): 230–44. doi:10.1093/oxrep/grh013

Everett, Helen, Shona Macleod, and Nalia Thurgood. 2013. *An Investigation of Head-teachers' and Teachers' Views Towards Science-Specific CPD.* Slough, UK: National Foundation for Educational Research (NfER), http://www.nfer.ac.uk/publications/SLCS01/SLCS01.pdf.

Goe, Laura. 2007. *The Link Between Teacher Quality and Student Outcomes: A Research Analysis.* Washington, DC: National Comprehensive Center for Teacher Quality, http://www.niusileadscape.org/docs/FINAL_PRODUCTS/LearningCarousel/LinkBetweenTQandStudentOutcomes.pdf.

Hanushek, Eric A., and Ludger Wößmann. 2007. *The Role of Education Quality in Economic Growth* (World Bank Policy Research Working Paper 4122). Washington, DC: World Bank, https://openknowledge.worldbank.org/bitstream/handle/10986/7154/wps4122.pdf.

Harris, Douglas N., and Tim R. Sass. 2011. "Teacher Training, Teacher Quality and Student Achievement." *Journal of Public Economics* 95 (7/8): 798–812. doi:10.1016/j.jpubeco.2010.11.009

Hattie, John. 2003. "Teachers Make a Difference: What Is the Research Evidence?" Paper presented at the Australian Council for Educational Research Annual Conference 2003, Sydney, Australia, http://www.decd.sa.gov.au/limestonecoast /files/pages/new% 20page /PLC/teachers_make_a_difference.pdf.

Ingersoll, Richard M. 2007. *A Comparative Study of Teacher Preparation and Qualifications in Six Nations* (CPRE Policy Briefs RB-47). Philadelphia, PA: Consortium for Policy Research in Education, University of Pennsylvania, http://repository.upenn .edu/cgi/ viewcontent.cgi?article=1145&context=gse_pubs.

Ingvarson, Lawrence, and Ken Rowe. 2008. "Conceptualising and Evaluating Teacher Quality: Substantive and Methodological Issues." *Australian Journal of Education* 52 (1): 5–35. doi:10.1177/000494410805200102

Mokhele, Matseliso. 2013. "Empowering Teachers: An Alternative Model for Professional Development in South Africa." *Journal of Social Science* 34 (1): 73–81.

UNESCO Institute for Statistics. 2006. *Teachers and Educational Quality: Monitoring Global Needs for 2015*. Montreal, Canada: Author, http://unesdoc.unesco.org /images/0014/001457/145754e.pdf.

Vlaardingerbroek, Barend, and Neil Taylor. 2014. *Issues in Upper Secondary Science Education: Comparative Perspectives*. New York, NY: Palgrave Macmillan.

Whitehouse, Claire. 2011. *Effective Continuing Professional Development for Teachers*. Manchester, UK: Centre for Education Research and Policy, Assessment and Qualifications Alliance (AQA), https://cerp.aqa.org.uk/sites/default/ files/pdf_upload /CERP-RP-CW-19052011.pdf.

Further Reading

Akiba, Motoko, Gerald K. LeTendre, and Jan P. Scribner. 2007. "Teacher Quality, Opportunity Gap, and National Achievement in 46 Countries." *Educational Researcher* 36 (7): 365–87. doi:10.3102/0013189X07308739

Bolyard, Johnna J., and Patricia S. Moyer-Packenham. 2008. "A Review of the Literature on Mathematics and Science Teacher Quality." *Peabody Journal of Education* 83: 509–35. doi:10.1080/01619560802414890

Darling-Hammond, Linda. 2000. "How Teacher Education Matters." *Journal of Teacher Education* 51 (3): 166–73. doi:10.1177/0022487100051003002

Glewwe, Paul. 2006. "Schools, Teachers, and Education Outcomes in Developing Countries." In *Handbook of the Economics of Education*, vol. 2, edited by Eric Hanushek and Finis Welch, 945–1017. Oxford, UK: Elsevier.

Goe, Laura, and Leslie Stickler. 2008. *Teacher Quality and Student Achievement: Making the Most of Recent Research*. Washington, DC: National Comprehensive Center for Teacher Quality, http://files.eric.ed.gov/fulltext/ED520769.pdf.

2

Australia

Terry Lyons

Teaching Science in Australia

Backdrop

Science education has been the subject of increasing public interest over the last few years. While a good part of this attention has been due to the fundamental reshaping of school curricula and teacher professional standards currently underway, there has been a heightened level of critical media commentary about the state of science education in schools and science teacher education in universities. In some cases, the commentary has been informed by sound evidence and balanced perspectives. More recently, however, a greater degree of ignorance and misrepresentation has crept into the discourse.

Australia has a population of nearly 24 million people and a school population of around 3.6 million students. While it has a federal government, the six states (Queensland, New South Wales [NSW], Victoria, Tasmania, South Australia, and Western Australia) and two territories (Northern Territory and the Australian Capital Territory [ACT]) making up the federation are responsible for school education. Nevertheless, the overwhelming bulk of funding for school education is allocated to states and territories from federal government coffers, an arrangement that gives the federal government considerable and increasing leverage in shaping education policy.

About 71 percent of schools are administered directly by government departments of education, 18 percent of schools are Catholic schools, and 11 percent are independent schools, most of which also have religious affiliations (Australian Bureau of Statistics 2014). The relatively high proportion of non-government schools reflects their significant role in school education in Australia since European settlement in the late eighteenth century. It was only from the middle of the nineteenth century that state governments began to take greater responsibility for schools, with Victoria becoming the first state to provide free, secular, public school education in 1872. Non-government schools receive some funding from the federal government, are registered by state/territory governments, and must adhere to government-mandated curricula, standards, and reporting protocols.

School progression structures and requirements vary across states and territories. Generally, though, children begin school in a pre-primary year (designated kindergarten, reception, or foundation) at age four or five. Schooling is compulsory from Year 1 (age five or six) until Year 10 (age 15 or 16). In most states/territories, primary students graduate to secondary school at the end of Year 6 (ages 12 to 13). Of those starting secondary school, around 80 percent continue on to Years 11 and 12. With the exception of a few colleges catering exclusively for senior students, most junior and senior secondary students attend the same campus. The proportion of Year 10 students continuing to Year 11 has increased over the last decade or so, due in particular to legislation in 2010 requiring young people, unless employed, to remain at school or in other education or training until age 17. Around 86 percent of Year 10 females continue to Year 12 compared to 78 percent of males (Australian Bureau of Statistics 2014).

School Science

Science is a compulsory key learning area for all primary and secondary students up to Year 10. In primary schools, the time allocated to science lessons is generally between 1.5 to 2.5 hours per teaching week, or 6 to 10 percent of instructional time. In the past, primary science curricula varied across the country, but since the implementation of a national curriculum between 2012 and 2014, all primary students have been learning similar concepts and skills at similar stages. The Australian Curriculum: Science includes three strands:

- Science understanding, that is, knowledge and understanding of biology, chemistry, physics, and Earth and space science content;
- Science inquiry skills, for example, questioning, predicting, planning, and conducting experiments; and
- Science as a human endeavor, which covers the nature, history, usefulness, and influence of science.

The majority of primary school teachers are generalists without a science degree or science major, although recently there have been calls for every primary school to also have specialist teachers in science and mathematics (see, for example, Teacher Education Ministerial Advisory Group 2014).

The three strands extend into junior secondary school up to Year 10, thus constituting a continuum of learning. Science programs from Foundation to Year 10 are integrated rather than separate discipline subjects, although some schools organize their Year 10 teaching for a term or semester along discipline lines; for example, a term on biology topics followed by a term on chemistry topics. Running across the strands are three cross-curriculum priorities: sustainability, Aboriginal and Torres Strait Islander histories and cultures, and Australia's engagement with Asia. Embedded within the Australian Curriculum is a focus on seven general capabilities identified as national education goals: literacy, numeracy, information and communication technology capability, critical and creative thinking, ethical understanding, personal and social capability, and intercultural understanding. Science programs developed for primary and junior secondary years must integrate elements of all these strands, sub-strands, general capabilities, and cross-curriculum priorities.

The end of Year 10 once marked a significant juncture in a student's education, with external assessments and certificates of completion marking the end of compulsory schooling for many. The certificates of completion are still awarded, but with the increased retention into senior secondary this transition point has become less consequential for continuing students. Students progressing to Years 11 and 12 choose from a wide variety of subjects, with English the only compulsory subject in most states/territories.

The most popular science subjects in Year 12 are biology, chemistry, and physics, typically selected by around 25 percent, 18 percent, and 14 percent of students respectively. Around two percent of students choose Earth and environmental sciences, while four percent choose a multi-strand science subject. The proportion of senior students choosing science subjects has declined substantially over the last two decades. While the overall number of Year 12 students has increased by over 30,000 since 1992, the enrolments in physics, chemistry, and biology have decreased by around 8,000, 4,000, and 13,000 respectively (Kennedy, Lyons, and Quinn 2014).

At present there is no nationally consistent assessment or certification regime at the end of Year 12 and no concrete plans to introduce one. In most states and territories, Year 12 students undertake a combination of externally administered examinations and internal school assessments. The weighting of internal assessment varies from 50 percent in the case of NSW to 70 percent in South Australia and the Northern Territory to 100 percent in Queensland and the ACT. These last two have a system of moderated school-based assessment, although students considering tertiary entrance also sit an externally set test of university-suitability skills in the case of the ACT, or a core skills test in the case of Queensland. The Queensland model is slated for revision in 2015, with the new model likely to include an element of external assessment.

The various state/territory assessment regimes lead to an end-of-school credential such as the Higher School Certificate or the Certificate of Education, and can also help calculate a student's Australian Tertiary Admissions Ranking (ATAR) or Overall Position (OP), which individual universities then use to regulate entry to different courses. The ATAR and OP are rankings of a school leaver's position relative to other students in the cohort based on aggregated assessment scores in qualifying subjects.

Until very recently, state and territory boards of study tended to develop their own senior curricula independently of one another. However, a number of subjects developed for the Australian Curriculum, including physics, chemistry, biology, and Earth and environmental science, were endorsed by all state and territory education ministers in 2012 as a common basis for new state/territory curricula to be implemented over the next few years.

Teacher Academic and Professional Education and Training

Historical Overview

Teacher preparation in nineteenth-century Australia consisted of a four-month apprenticeship-type course in a model school. From the early twentieth century until the beginning of the 1960s, teacher training was primarily the responsibility of state-controlled teachers' colleges staffed by experienced teachers who focused on practical training for the classroom (Barcan 1995). In addition, a comparatively small number of secondary teachers qualified via a Bachelor's degree followed by a one-year Diploma of Education (Committee on the Future of Tertiary Education in Australia 1964). Beginning teachers served a probation period after which they were inspected; if deemed competent, they received their registration. A chronic shortage of secondary teachers between the 1940s and 1970s led to downward pressure on entry standards. This was offset in part by a paucity of alternative jobs for academically educated people, resulting in a teacher cohort that Barcan (1995, 50) described as being of "fair average quality."

In the late 1960s and early 1970s, the teacher college system was subsumed within newly formed colleges of advanced education (CAEs) responsible for post-secondary education in teaching, nursing, accountancy, and a number of other technical fields. Undergraduate teaching qualifications lengthened to three years for primary and junior secondary teachers, and gradually more universities began to offer secondary teacher training courses. Barcan (1995) reports that the curriculum expanded to include more educational theory, while the backgrounds of university and college lecturers included less classroom experience. The National Inquiry into Teacher Education (Auchmuty 1980) and other state-based inquiries advocated a transition to a four-year teacher education program for secondary teachers and justified the rapid growth of specialist provision in mathematics and science (Knight, Lingard, and Bartlett 1994). During this period, there was greater pressure for teacher educators to attain higher degree qualifications, including doctorates.

In line with the Dawkins higher-education reforms (Dawkins 1987) of the late 1980s and early 1990s, CAEs amalgamated with universities, giving rise to the current arrangement whereby secondary teacher education is the preserve of faculties or schools of education at 37 universities and six non-university colleges.

The actual governmental responsibility for teacher education is unclear. While nominally this had long been the responsibility of individual states, Chadbourne (1997) described it as a de facto national system because universities are funded by the Commonwealth. These circumstances have, however, resulted in a national teacher education system that is reasonably homogeneous.

Classroom teachers, regardless of whether they work in government, Catholic, or independent schools, are accredited by teacher registration bodies in each state/territory with reference to the professional standards for teachers developed by the Australian Institute of Teaching and School Leadership (AITSL). These standards were informed by and distilled from many of the standards that various state and territory bodies explicitly or implicitly expected of teachers and school leaders. The AITSL standards constitute agreed expectations of teachers' capabilities at four career stages—graduate, proficient, highly accomplished, and lead. Capabilities encompass teachers' professional knowledge of their discipline content and their students, their professional practice in planning, teaching, and assessing their students, and their professional engagement with current research and with education stakeholders (AITSL 2011).

Qualification Profile of Science Teachers Currently in Schools
Science teachers represent around 18 percent of all secondary teachers and constitute the third largest group of secondary teachers after English and mathematics teachers (McKenzie et al. 2014). Most qualified science teachers in Year 7–12 schools teach senior and junior classes. Due to the integrated science curriculum in the junior secondary years, teachers must be equipped to teach across a broad range of physics, chemistry, biology, and Earth and environmental science topics at this level, as well as having depth in their specific discipline area (e.g., chemistry) in the senior years. Hence, a chemistry teacher will usually have (at least) two science teacher identities—the generalist science teacher and the senior chemistry teacher. In addition, there has been a national move toward requiring secondary teachers to be qualified to teach in two different subject areas, for example physics and mathematics, or biology and physical education. AITSL now requires secondary initial teacher education (ITE) programs to provide at least a major study in one teaching area and preferably a second teaching area comprising at least a minor study.

The ratio of graduate to undergraduate trained science teachers differs considerably from the 50:50 found among secondary teachers more generally. Around 62 percent of current science teachers qualified via a 12-month or 18-month Graduate Diploma of Education (GDED) subsequent to their three-year undergraduate science degrees. This path tends to have been favored more by physics teachers, 72 percent of whom qualified this way (Weldon et al. 2014). The remaining 38 percent of science teachers qualified via an undergraduate ITE program in which they studied a combination of science discipline units, science curriculum units, and general education units. The most common of these is a four-year Bachelor of Education, although some teachers hold a combined or double degree such as a Bachelor of Teaching/ Bachelor of Science. While some older teachers would have originally qualified via a three-year Bachelor of Teaching or Diploma of Teaching, most of these teachers have subsequently completed a fourth year to upgrade their qualification.

Qualification Requirements for Current Preservice Science Education Students

Whereas most undergraduate science education students begin their ITE courses soon after completing Year 12, a substantial proportion of the GDED teachers come into teaching from careers in science, engineering, or associated fields. According to McKenzie et al. (2014), the graduate ITE path has recently increased in popularity. Anecdotal evidence suggests this increase may be due in part to prospective teachers taking the opportunity to enroll in GDED courses before these are phased out and replaced by a two-year minimum Master's-level ITE course as advocated by AITSL and outlined below.

The teacher education landscape is in the process of being reshaped, with all stakeholders—federal and state government and non-government education bodies, universities, schools, professional associations, teacher registration boards, and the like—gradually revising their programs, expectations, and cultures with reference to the AITSL professional standards for teachers and the Australian Curriculum. Universities in particular have had to restructure their undergraduate and graduate ITE programs in order to comply with AITSL and state/territory teacher registration requirements.

With respect to generic learning skills, the expectation is for entrants to ITEs to have the capacity to engage in "a rigorous higher education program and to carry out the intellectual demands of teaching itself" (AITSL 2013, C4) and to have levels of literacy and numeracy broadly equivalent to those of the top 30 percent of the population. With respect to subject content knowledge, AITSL requires, as noted earlier, undergraduate secondary ITE programs to comprise a major study in at least one teaching area. In most programs, this study area equates to six units of science, with no more than two units at the first-year level and no fewer than two units at the third-year level. AITSL also recommends inclusion of a minor study (which is actually now compulsory in some states), usually of four units but with no more than two at the first-year level.

In terms of professional studies in education, AITSL requires undergraduate ITE programs to include a minimum of two units of discipline-specific curriculum and pedagogical studies for each teaching area, as well as general education studies and 80 days of supervised professional experience in schools. The studies and school experience are deemed sufficient to prepare graduates to teach across all years of secondary schooling.

Entry into graduate-science ITE courses requires students to have already completed sufficient discipline units in their primary and secondary teaching areas during their undergraduate degrees. The graduate ITE course must comprise at least two years of full-time-equivalent professional studies in education. This includes a minimum of two units of discipline-specific curriculum and pedagogical studies for each teaching area, as well as general education studies and 60 days of supervised professional experience in schools. The graduate program should be such that it prepares beginning teachers to teach junior and senior students.

Continuing Professional Development

The rationale for, and types of, continuing professional development have been made more explicit and nationally consistent since the establishment of the AITSL professional standards for teachers, which provide a framework for career progression and promotion. The key standard in this respect requires teachers to engage in professional learning and improve practice, with learning content containing elaborations and illustrations of what is expected of teachers at each career stage (AITSL 2011).

The three most common professional learning topics that secondary science teachers undertake are how to use information and communication technologies (ICT) effectively, how to identify and use different teaching resources, and how to evaluate and improve teaching (Weldon et al. 2014, 51). In terms of science teachers' perceptions, secondary teachers as a whole appear to be less positive than primary teachers about the benefits of their professional learning opportunities, while secondary science teachers are less positive than other secondary teachers. The exception to this is professional learning focusing on the effective use of ICT, which teachers in all subject areas find equally beneficial.

Science Teacher Professional Associations

The Australian Science Teachers Association (ASTA) is the overarching body representing science teachers in primary and secondary schools. ASTA is a federation of eight state and territory science teacher associations, and it has around 4000 individual members. The role of ASTA and its state/territory affiliates includes the following:

- Encouraging advocacy for science education and science teachers in public policy;
- Forming links with stakeholders in government, education, and industry;
- Facilitating support and professional development for science teachers; and
- Providing science teachers with a forum for communication and discussion.

Governance is provided by the Federal Council with representation from state and territory associations, and the organization is administered nationally by a chief executive officer. ASTA hosts an annual science teachers' conference (CONASTA) and publishes the national journal *Teaching Science*.

A number of the state/territory associations also convene conferences and produce publications. The Australasian Science Education Research Association (ASERA) likewise includes teachers, but its membership extends more to science teacher educators and researchers. ASERA publishes the highly regarded journal *Research in Science Education* and hosts an annual international conference.

Many individuals have contributed to the robust, supportive, and productive science education associations in Australia, and it not the purpose of this chapter to detail their contributions. However, it is difficult to discuss the significant role of these organizations in the science education community without acknowledging the contributions of Peter Fensham. Peter was the first professor of science education in Australia and perhaps the first in the world outside the USA (Gunstone 2009). He was the first national president of ASTA, and he founded ASERA in 1970, making it the first professional body in educational research in the country. He has directly and indirectly shaped the development of school science curricula across the country and has arguably done more than anyone else to generate international recognition of Australian science education and research.

Issues in Upper Secondary Science Teacher Quality

Teacher Supply

Warnings about a looming shortage of qualified science teachers expressed more than a decade ago (see, for example, Rennie, Goodrum, and Hackling 2001; Ministerial Council on Education, Employment, Training and Youth Affairs 2003) are now being realized (Productivity Commission 2012), and the impending retirement of many physics and chemistry teachers will exacerbate this problem. Around 40 percent of physics teachers, 35 percent of chemistry teachers, and 24 percent of biology teachers are older than 51 years of age (Weldon et al. 2014). At an average age of nearly 47 years, physics teachers comprise the oldest cohort of all secondary teachers. The fact that the average age of physics and chemistry teachers increased by two years between the 2010 and 2013 Staff in Australian Schools surveys indicates that the supply of new teachers in these subject areas has failed to keep up with demand.

The supply problem appears to be worse in Australia than in most other countries, with an OECD study reporting that Australian students are more likely than students in many other countries to be enrolled in schools with a lack of mathematics and science teachers (OECD 2012). Likewise, the 2009 Programme for International Student Assessment (PISA) reported that around 24 percent of 15-year-old Australian students were enrolled in schools where the lack of qualified science teachers hindered instruction. This percentage compares unfavorably with the average across the participating PISA countries of 18 percent (Productivity Commission 2012). Shortages of science teachers are chronic in rural and remote Australia, with one national report concluding that secondary science teachers in rural areas are about twice as likely—and those in remote areas about four times as likely—as those in metropolitan areas to be working in a school in which it is very difficult to fill vacant science teaching positions (Lyons et al. 2006).

The shortages have led to a substantial amount of what is loosely referred to as out-of-field teaching, although the extent of this across Australia is difficult to determine and its impact on learning outcomes doubly so. Common examples of out-of-field teaching include biology teachers teaching chemistry and physical education teachers teaching biology. One difficulty in determining the extent of this practice is the lack of agreement about what qualifications constitute "in field," with stakeholders in this area tending to have different expectations. Weldon et al. (2014, 26), for example, concluded that there is "relatively little out-of-field teaching in these areas" based on their findings that a high proportion of the secondary teachers teaching biology, chemistry, general science, and physics have undertaken at least two years tertiary study in the area or tertiary training in teaching methodology in that field. Supporting this view, the OECD's Teaching and Learning International Survey (TALIS) of junior high school science teachers reported that only 5.6 percent of teachers in the Australian sample were underqualified for teaching science at that level (OECD 2014, 281). This figure compares favorably with the OECD international average of 7.6 percent and is lower than the national averages in a number of countries that tend to outperform Australian students in PISA, including Finland (10.5 percent), Japan (6.9 percent), and Korea (8.3 percent) (OECD 2014).

Other stakeholders, including the Australian Council of Deans of Science, consider this standard of qualification to be sub-optimal, preferring that all science teachers hold a full science degree with a major in their main science teaching subject (see, in this regard, Harris, Jensz, and Baldwin 2005). However, given the recent AITSL requirement of a minimum two-year teacher qualification, there is now some concern that a pathway requiring five years of study will further hinder the supply of science teachers.

A second issue is the lack of consensus about the impact of underqualified or out-of-field teachers on student learning outcomes. Some commentators consider it to be self-evident that students taught by a science teacher with a full science degree, a major in their subject, and appropriate teaching qualification will benefit from that expertise and outperform similar cohorts taught by teachers without a science degree. Nevertheless, evidence to support this contention is difficult to find. Year 8 students taking the 2011 Trends in International Mathematics and Science Study (TIMSS) science test performed similarly regardless of whether their teachers had a major in science, a major in education, or both. The authors of the 2011 TIMSS international report noted that "the issue of 'out-of-field' teaching appears not to be so much of a problem in science" (Thomson, Hillman, and Wernert 2012, 85). Similarly, a report by the Australian chief scientist concluded that there appears to be no definitive relationship between a country's mean science score in TIMSS or PISA and the qualifications of its teachers (Office of the Chief Scientist 2014).

However, this conclusion may not necessarily translate to senior classes, where one would expect that a deep rather than a broad discipline knowledge would be of greater importance. Unfortunately, insufficient Australian research has been undertaken to investigate relationships between science teacher qualifications and senior students' achievement.

Professional Education and Training Issues

Several national and localized programs have been initiated in response to concerns about supply and qualifications of science teachers. At the national level, the 2012 federal budget included a $54 million program directed toward school mathematics and science, including teacher education, and set up in response to the chief scientist's Mathematics, Engineering and Science in the National Interest report (Office of the Chief Scientist 2012).

One initiative that emerged from this funding was the Enhancing the Training of Mathematics and Science Teachers Programme (ETMSTP), which aims to improve the quality of teaching through more relevant and effective ITE programs and professional development, as well as by increasing the quantity of subject-qualified teachers. Five of the projects funded under this initiative have focused on developing sustained collaborations between education and science faculties in order to achieve a better integration of the courses and academic expertise involved in science teacher preparation. Examples of project approaches include the introduction of "taster" science/mathematics education and communication units into undergraduate science degrees, greater collaboration between research scientists and teacher educators in preservice science teacher education, and a greater focus on practical school experience. While these projects are still in development, the intention is that they will produce improved, evidence-based models that can be scaled up to have a national impact on science teacher preparation.

Such initiatives need to overcome a number of impediments if they are to generate greater interest in science teaching careers. First, the teaching profession is not held in particularly high status in Australia compared to countries such as China, Korea, Finland, and Japan (Marginson et al. 2013). According to the TALIS report (OECD 2014), only 38 percent of Australian lower secondary teachers agreed that society values the teaching profession. While this percentage was higher than the international mean of 32 percent, it was below the means of a number of countries—Canada, Finland, Singapore, and Korea among them—that tend to outperform Australia in international science tests. The relatively low status of the profession in Australia is reflected in entry requirements, which are not as competitive as those in some of the higher-performing countries (Marginson et al. 2013).

Second, the teaching profession in Australia is often criticized for a wide range of educational and societal deficiencies, which also regularly come under government and media scrutiny. According to Parkes (2013, 113), preservice teacher education "has always been an inherently political exercise, and appears to be an enduring public policy problem in Australia." A review by May et al. (2009) found that since 1965 there have been 146 state or federal government reviews, reports, and official statements concerning investigations into, or attempts to reform, teacher education. Many of the recent media headlines about school science referred to in the introduction to this chapter have tended to paint a somewhat negative picture of science teachers (see, for example, Ferrari 2014; McDougall 2014).

Then there are the challenges of the job itself. Quite apart from the demands of planning, teaching, and assessing, secondary school teaching is justifiably regarded as a personally challenging occupation. The TALIS report suggested that Australian classrooms are among the most challenging cross-nationally in terms of some types of student behavior. Around 10 percent of junior high school teachers reported staff being intimidated or verbally abused; this percentage was the third highest among the 34 countries that participated in the study (OECD 2014). For prospective science teachers, such disadvantages are not necessarily compensated by remuneration. Milanowski (2003), moreover, reports that highly skilled graduate science teachers are those teachers who persist for the least time in teaching because they can earn higher salaries in other occupations.

The gender profile of physics teachers and, to a lesser extent, chemistry teachers reveals a further challenge in attracting young people to these fields. While the ratio of males to females in general science is around 49:51, around 77 percent of physics teachers and 58 percent of chemistry teachers are male. Despite a number of initiatives to encourage females into physics and physics teaching, the proportion of male physics teachers has increased since 2007, tending to further reinforce the classic stereotype of physicists as older males. This pattern may have implications for student perceptions of and aspirations to engage in physics teaching.

Issues at the Chalkface

Secondary science teachers work around 46 hours per week on school-related activities, a figure similar to that of teachers in other curriculum areas (Weldon et al. 2014). Of these hours, 20 are spent teaching, 7 hours planning, and 5 hours marking, which was around the average for the countries participating in TALIS (OECD 2014). Like most teachers in this study, Australian senior secondary teachers overwhelmingly claimed to embrace constructivist principles in the classroom, with 93 percent agreeing that a teacher's role is to facilitate the student's own inquiry. In most other respects, the practices of Australian teachers are similar to those of their colleagues in the majority of TALIS countries. However, the study also indicates that Australian teachers have a greater tendency toward collaboration, particularly with respect to sharing teaching materials, discussing student issues with colleagues, and seeking consensus about evaluation standards.

Research shows that the quality of science laboratory facilities varies considerably from school to school in Australia. A national study by Hackling (2009) found 54 percent of schools rated their science teaching facilities as good or very good, while 15 percent rated them as poor or very poor. There is some indication that both the availability of and support for laboratory technicians require attention. According to Hackling, around 40 percent of schools have difficulty recruiting technicians. One of the main impediments to attracting suitable applicants is "poor conditions of service, in particular the poor match between salary levels and responsibility" (2009, 4). Consistent with this variability, while 70 percent of schools regarded the level of technical support as good or very good, around 10 percent indicated it was poor or very poor. Given around 40 percent of technicians in Hackling's study were aged 50 or over, there are worries that recruitment will become increasingly difficult unless employment conditions are improved. Of the concerns the laboratory technicians expressed, lack of opportunity for ongoing training came in for particularly strong attention; around one in five technicians indicated that they needed further support or training to competently perform many of their primary tasks. In particular, around a quarter reported a need for further support or training in regard to a number of fundamental safety issues such as first aid, accident and emergency procedures, fire extinguishers, disposal of hazardous waste, and preparation of risk assessment sheets (2009, 44).

Continuing Professional Development Issues

Science teachers tend to undertake less professional learning and to find it less effective than do their colleagues in other subject areas. According to research conducted by Weldon et al. (2014), general science teachers averaged just over 8.0 days a year in professional learning activities. Senior physics, chemistry, and biology teachers spent between 7.0 and 7.5 days, below the national average of 8.2 days for all secondary teachers. In terms of access to professional learning, Weldon et al. (2014) reported that only 56 to 59 percent of science teachers (depending on the teaching area) had undertaken any in the preceding 12 months. By comparison, 79 percent of English teachers, 74 percent of mathematics teachers, and 68 percent of information technology teachers had accessed at least one professional learning opportunity during that period.

Australia's forthcoming curriculum revisions make it imperative that the country's senior science teachers are provided with and encouraged to attend targeted and effective professional learning opportunities. However, as Marginson et al. (2013) note, no significant project has been put in place to develop the professional learning that teachers need if they are to support the changed practices implied in the curriculum document.

It should also be noted that science teachers in rural and remote schools tend to have a greater unmet need than their colleagues in metropolitan centers for involvement in curriculum development (Lyons et al. 2006). It is therefore important that particular efforts be made to ensure rural and remote voices are heard in this process, so that metro-centric curriculum content and contexts are not the only ones represented, especially in the "science as a human endeavor" strand of the curriculum.

Trends and Developments in Upper Secondary
Science Teacher Quality

Science curricula and assessment in the senior years are currently undergoing a period of substantial restructure and refocus. As noted, education ministers in each state and territory have endorsed the new Australian curricula for senior biology, chemistry, physics, and Earth and environmental science as a common basis for the development of their own state/territory courses. Some states, such as Western Australia, introduced a suite of new courses in 2015, while others, such as NSW, have yet to specify an implementation time frame.

Despite these developments, plans have been particularly tentative since 2013 when the incoming federal government convened a review of the Australian Curriculum and the state/territory boards of study subsequently postponed any curriculum development until they could digest the implications of the review findings. Senior science teachers therefore likely face an intense period of curriculum development, evaluation, and revision over the next few years, along with the uncertainty that comes with extensive consultation processes. The senior science curriculum is a contested field at the best of times, with a range of stakeholders advocating their particular positions.

Even though the consultation period for the national curriculum has concluded (notwithstanding any significant changes as a result of the aforementioned review), the next few years will prove to be particularly unsettling for teachers as they navigate the machinations of state/territory curriculum development processes. With respect to any federal aspirations toward a consistent national assessment and accreditation regime, however, education ministers at the December 2012 meeting asserted their rights to continue state/territory responsibilities for certification, assessment, and examination requirements.

Another issue currently in the spotlight—and likely to remain there for some time—is the variation in entry requirements for ITE courses. Currently, there is no mandatory government minimum ATAR or OP for entry into teacher education courses. While AITSL and state/territory teacher registration bodies are able to influence teacher quality via course accreditation and teacher standards, they do not determine the requirements for entry into undergraduate ITEs. Rather, universities and other institutes apply their own entry criteria, including using ATAR or OP scores as the basis of setting their entry cut-off points for different courses.

Entry to high-demand or high-status courses such as medicine at prestigious universities tends to require the top ATAR or OP rankings. However, the practice of universities setting ATAR entry requirements for ITE courses each year has led to significant variations not only between institutions but from year to year within the same institution, depending to some extent on internal calculations of enrolment quotas. For example, a high-ranking, high-demand city university may set an ATAR of 75 for entry to an ITE science course, while a lower-demand regional or multi-campus university may set an entry ATAR of 55 or under (Preiss and Butt 2013). In some cases, entry cut-offs for the same program can be different at separate campuses of the same institution. Some universities have been accused of allowing substantial numbers of applicants to enroll in ITE courses without having achieved the published ATAR or OP cut-off. Indeed, AITSL (2014) reported that between 2005 and 2012, the percentage of students from the lower ATAR bands entering teacher education had increasing substantially. Such practices have prompted criticisms that entry standards are subject to the vagaries of an increasingly competitive tertiary environment and are neither fair nor transparent.

In February 2014, the federal government established the Teacher Education Ministerial Advisory Group (TEMAG) to provide advice on "how teacher education programmes could be improved to better prepare new teachers with the practical skills needed for the classroom" (TEMAG 2014, 52). The purpose and constitution of the group generated some controversy among those advocating higher entry standards in teacher education. Several of these people were quick to point out that the chair of the group is also vice-chancellor at one of the universities with the lowest ITE entry requirements in the country.

The TEMAG report, released at the end of 2014, made a number of recommendations, including the establishment of a national ITE regulator to oversee a more rigorous accreditation process whereby ITE institutions must demonstrate both the quality of their programs and student achievement of published graduate outcomes. The committee balked, however, at proposing minimum entry standards based on ATAR or OP rankings, preferring instead to recommend that providers improve the transparency of their ITE entry criteria and practices.

The current and anticipated revisions to important domains of science teaching and science teacher education—new senior curricula, recent AITSL graduate standards, the chief scientist's STEM teacher education reform agenda, and implementation of TEMAG recommendations—all signal a dynamic period for science teacher education in Australia. Aspiring teachers are either hurrying to enroll in the few remaining short graduate courses before these expire or considering whether they have the capacity to undertake one of the new two-year Master's-level courses. Existing science teachers are considering the best approaches to implementing new curricula and identifying which professional development opportunities will best support them. Preservice teacher educators who have only just revised their courses to comply with AITSL graduate teacher standards will soon need to do so again in light of revisions to the Australian Curriculum and the aforementioned assessment changes. Many science and education faculty academics involved in the ETMST program are developing new channels of communication and collaboration to explore how they can realize a shared vision for science teacher education; they are also contemplating how to encourage more of their students to consider a teaching career. The outcomes of these initiatives at this point are uncertain, but the degree of activity in the field is at least testament to the level of energy and commitment in the system to improving science teacher education.

References

Auchmuty, James. 1980. *Report of the National Inquiry into Teacher Education (Australia)* (the Auchmuty report). Canberra, ACT: AGPS.

Australian Bureau of Statistics. 2014. *Schools, Australia, 2013* (ABS catalogue number 4221.0). Canberra, ACT: Author, http://www.abs.gov.au/AUSSTATS/abs@.nsf/Lookup/4221.0Explanatory%20Notes12013?OpenDocument.

Australian Institute of Teaching and School Leadership (AITSL). 2011. *Australian Professional Standards for Teachers*. Melbourne, VIC: Education Services Australia, http://www.aitsl.edu.au/australian-professional-standards-for-teachers/standards/list.

———. 2013. *Accreditation of Initial Teacher Education Programs in Australia: Guide to the Accreditation Process*. Carlton South, VIC: Education Services Australia, http://www.aitsl.edu.au/docs/ default-source/default-document-library/guide_to_the _accreditation_process.pdf? sfvrsn=4.

———. 2014. *Initial Teacher Education: Data Report 2014*. Melbourne, VIC: Author, http://www.aitsl. edu.au/ initial-teacher-education/data-report-2014.

Barcan, Alan. 1995. "The Struggle over Teacher Training." *Agenda* 21 (1): 49–62.

Chadbourne, Rod. 1997. "Teacher Education in Australia: What Difference Does a New Government Make?" *Journal of Education for Teaching* 23 (1): 7–27.

Committee on the Future of Tertiary Education in Australia. 1964. *Tertiary Education in Australia: Report to the Australian Universities Commission* (the Martin report), vol. 1. Melbourne, VIC: Commonwealth of Australia.

Dawkins, John. 1987. *Higher Education: A Policy Discussion Paper*. Canberra, ACT: AGPS.

Ferrari, Justine. 2014. "Quality of Teaching at the Heart of Education Problems." *The Australian*, 19 July.

Gunstone, Richard. 2009. "Peter Fensham: Head, Heart and Hands." *Cultural Studies of Science Education* 4: 303–14.

Hackling, Mark. 2009. *The Status of School Science Laboratory Technicians in Australian Secondary Schools: Research Report Prepared for the Department of Education, Employment and Workplace Relations, Edith Cowan University, Perth*. Mount Lawley, WA: Edith Cowan University.

Harris, Kerri-Lee, Felicity Jensz, and Gabrielle Baldwin. 2005. *Who's Teaching Science? Meeting the Demand for Qualified Science Teachers in Australian Secondary Schools. Report Prepared for Australian Council of Deans of Science*. Melbourne, VIC: Centre for the Study of Higher Education, University of Melbourne.

Kennedy, John, Terry Lyons, and Frances Quinn. 2014. "The Continuing Decline of Science and Mathematics Enrolments in Australian High Schools." *Teaching Science*, 60 (2): 34–46.

Knight, John, Bob Lingard, and Leo Bartlett, L. 1994. "Reforming Teacher Education Policy Under the Labor Government in Australia 1983–93." *British Journal of Sociology of Education: Special Issue. Teacher Education: Past, Present and Future* 15 (4): 451–66.

Lyons, Terry, Ray Cooksey, Debra Panizzon, Anne Parnell, and John Pegg. 2006. *Science, ICT and Mathematics Education in Rural and Regional Australia: The SiMERR National Survey*. Canberra, ACT: Department of Education, Science and Training, http://simerr.une.edu.au/pages/projects/1nationalsurvey/Abridged%20report /Abridged_Full.pdf.

Marginson, Simon, Russell Tytler, Brigid Freeman, and Kelly Roberts. 2013. *STEM: Country Comparisons: International Comparisons of Science, Technology,*

Engineering and Mathematics (STEM) Education. Final Report. Melbourne, VIC: Australian Council of Learned Academies, www.acola.org.au.

May, Josephine, Allyson Holbrook, Alison Brown, Greg Preston, and Bob Bessant. 2009. *Claiming a Voice: The First Thirty-Five Years of the Australian Teacher Education Association.* Perth, WA: The Australian Teacher Education Association, http://hdl.handle.net/1959.13/917676.

McDougall, Bruce. 2014. "Back to School for Our Bad Teachers" (later amended to "Improving Standards: Back to School for Our Teachers"). *The Daily Telegraph,* November 20, 2014.

McKenzie, Philip, Ruth Weldon, Glenn Rowley, Martin Murphy, and Julie McMillan. 2014. *Staff in Australia's Schools 2013: Main Report on the Survey.* Camberwell, VIC: Australian Council for Educational Research, http://research.acer.edu.au/tll_misc/20.

Milanowski, Anthony. 2003. "An Exploration of the Pay Levels Needed to Attract Students with Mathematics, Science and Technology Skills to a Career in K–12 Teaching." *Education Policy Analysis Archives,* 11, http://epaa.asu.edu/ojs/article/view/278.

Ministerial Council on Education, Employment, Training and Youth Affairs (MCEETYA). 2003. *Demand and Supply of Primary and Secondary School Teachers in Australia.* Melbourne, VIC: Ministerial Council on Education, Employment, Training and Youth Affairs.

Office of the Chief Scientist. 2012. *Mathematics, Engineering and Science in the National Interest.* Canberra, ACT: Department of Industry, Innovation, Science, Research and Tertiary Education, http://www.chiefscientist.gov.au/wp-content/uploads/Office-of-the-Chief-Scientist-MES-Report-8-May-2012.pdf.

———. 2014. *Benchmarking Australian Science, Technology, Engineering and Mathematics.* Canberra, ACT: Commonwealth of Australia, http://www.chiefscientist.gov.au/wp-content/uploads/BenchmarkingAustralianSTEM_Web_Nov2014.pdf.

Organisation for Economic Co-operation and Development (OECD). 2012. *Preparing Teachers and Developing School Leaders for the 21st Century: Lessons from around the World.* Paris: OECD Publishing.

———. 2014. *TALIS 2013 Results: An International Perspective on Teaching and Learning.* Paris: OECD Publishing, http://dx.doi.org/10.1787/9789264196261-en.

Parkes, Robert, J. 2013. "Challenges for Curriculum Leadership in Contemporary Teacher Education." *Australian Journal of Teacher Education,* 38 (7): 112–28.

Preiss, Benjamin, and Craig Butt. 2013. "Teacher Entry Ranking Tumbles." *The Age,* January 18, 23, http://www.theage.com.au/data-point/teacher-entry-ranking-tumbles-20130117-2cwb5.html.

Productivity Commission. 2012. *Schools Workforce* (research report). Canberra, ACT: Commonwealth of Australia, http://www.pc.gov.au/inquiries/completed/education-workforce-schools/report/schools-workforce.pdf.

Rennie, Leonie, Denis Goodrum, and Mark Hackling. 2001. "Science Teaching and Learning in Australian Schools: Results of a National Study." *Research in Science Education* 31: 455–98. doi:10.1023/A:1013171905815

Teacher Education Ministerial Advisory Group (TEMAG). 2014. *Action Now: Classroom Ready Teachers.* Canberra, ACT: Department of Education, http://www.studentsfirst.gov.au/teacher-education-ministerial-advisory-group.

Thomson, Sue, Kylie Hillman, and Nicole Wernert. 2012. *Monitoring Australian Year 8 Student Achievement Internationally: TIMSS and PIRLS 2011.* Camberwell, VIC: Australian Council for Educational Research.

Watt, Helen M., and Paul W. Richardson. 2008. "Motivations, Perceptions, and Aspirations Concerning Teaching as a Career for Different Types of Beginning Teachers." *Learning and Instruction* 18: 408–28. doi:10:1016/j.learninstruc.2008.06.002

Weldon, Paul, Julie McMillan, Glenn Rowley, and Phillip McKenzie. 2014. *Profiles of Teachers in Selected Curriculum Areas: Further Analyses of the Staff in Australia's Schools 2013 Survey.* Camberwell, VIC: Australian Council for Educational Research, under contract with Commonwealth of Australia, https://docs.education.gov.au/node/36281.

Further Reading

Australian Institute of Teaching and School Leadership (AITSL). *Accredited Programs List.* Melbourne, VIC: Author, http://www.aitsl.edu.au/initial-teacher-education/accredited-programs-list.

Lyons, Terry. 2014. "Australia." In *Issues in Upper Secondary Science Education: Comparative Perspectives,* edited by Barend Vlaardingerbroek and Neil Taylor, 11–32. New York: Palgrave Macmillan.

Thomson, Sue, Kylie Hillman, Nicole Wernert, Marina Schmid, Sarah Buckley, and Ann Munene. 2012. *Monitoring Australian Year 4 Student Achievement Internationally: TIMSS and PIRLS 2011.* Camberwell, VIC: Australian Council for Educational Research.

3

Singapore

Lee Chin Chew and Kim Chwee Daniel Tan

Teaching Science in Singapore

Backdrop

Singapore, a former British colony, became an independent republic in 1965 when it separated from the Federation of Malaysia, with which, as a constituent state, Singapore had previously achieved independence from Britain. Singapore then developed, as popularly described, "from being a third world to becoming a first world country" in just 50 years. Singapore's "miraculous" change into a modern city-state involved rapid economic growth as well as intensive modernization.

The city-state's multi-ethnic population presently stands at 5.47 million, with Chinese comprising the majority ethnicity at 74 percent of the resident population. The indigenous Malays comprise 13 percent of the resident population, Indians 9.1 percent, and others (Eurasians, etc.) 3.9 percent (Department of Statistics Singapore 2014).

The dominance of one political party in parliament during these developmental years means that Singapore's government has played a direct and active role through its various ministries in formulating and implementing policies. The Ministry of Education, for instance, has had total charge of Singapore's education policies, which through the years have been driven by the need to gel the various ethnic factions into one nation even as the overwhelming priority has been to educate the city-state's people to meet the manpower needs of developing industries.

Children start formal education in national schools at the age of seven, after several years of optional pre-primary learning (Ministry of Education Singapore 2014c). In 2013, a total of 237,000 children were enrolled in 182 primary schools. Students undergo six years of primary schooling and then proceed to four or five years of secondary education. At the end of Primary 6, they take the Primary School Leaving Examination (PSLE), which assesses their suitability for one of three secondary-level streams—the express stream, the normal (academic) stream, and the normal (technical) stream. Of the 48,000 Secondary 1 students (equivalent to Grade 7) in the 2013 cohort, 60 percent entered the four-year express stream, 26.5 percent the five-year normal (academic) stream, and the rest the four-year normal (technical) stream (Ministry of Education 2014a).

Singaporean education has increasingly been rendered more flexible and diverse so as to provide students with a wider choice of educational institutions and thus opportunities to develop diverse strengths and interests (Ministry of Education Singapore 2014b). More choices of school types at the secondary level are also now available. "Independent," "independent specialized," "mixed-level," and "specialized schools" have also been established alongside the existing government and government-aided schools. While specialized schools offer customized programs for students who are inclined toward hands-on and practical learning, the specialized independent schools offer programs designed to extend talented students in areas such as mathematics, the sciences, the arts, and sports.

In 2013, a total of 178,000 students were in 154 secondary schools, and 37,000 in 15 mixed-level schools. Mixed-level schools offer a six-year integrated program for academically strong students who prefer a more independent and less structured learning style. The program extends beyond secondary schooling to include two years of pre-university learning. It differs from those programs where students end their secondary schooling after obtaining their General Certificate in Education, Ordinary-level (GCE O-level), and then opt for post-secondary programs, which for many will be two- or three-year pre-university studies leading to GCE Advanced-level (A-level). Mixed-level schooling offers a continuous learning program leading directly to GCE A-level and thus bypassing the GCE O-level.

About 30 percent of students continue with two or three years of pre-university schooling after finishing their secondary schooling. In 2013, approximately 22,000 students were enrolled in 14 pre-university institutions (called "junior colleges"). Students who advance to pre-university education have to select a track (science, arts, or commerce) that leads to GCE A-level certification. Two national schools offer the International Baccalaureate (IB) higher-level program, which is an alternative qualification for university admission.

As alternatives to pre-university education, students can opt for post-secondary routes such as vocational courses in a polytechnic or an institute of technical education. About 45 percent of a yearly cohort pursue polytechnic diplomas in the nation's five polytechnics, and 20 percent enroll in training courses in the institutes of technical education. However, a university education is much desired by the nation's young adults. There are six public universities in Singapore with a total student enrolment of about 60,000. In 2013, 16,000 students graduated with a first degree.

Crucial in an education system is the quality of teachers. Of the 14,800 teachers who teach at the primary level, 70 percent are university graduates, 9 percent hold Master's degrees, and less than 0.1 percent hold Doctoral degrees.

Of the 15,000 teachers who teach at the secondary levels, 94 percent are university graduates, 15 percent hold a Master's degree, and 0.3 percent hold Doctoral degrees. Of the 3,000 teachers at the pre-university level, almost all are university graduates; 22 percent hold Master's degrees, and 1.6 percent hold Doctoral degrees (Ministry of Education Singapore 2014a).

School Science

Singapore has long been committed to providing science education for its young people. From the time Singapore became an independent nation in 1965, the government embarked on a rigorous modernization program directed toward establishing a manufacturing economy and which, in recent years, has been tapping into high technology. The aim of the national school science curriculum has been to equip students with the knowledge and skills they need for entry into the workforce. An important inclusion is training in the computer technology skills that enable young people to excel in today's technology-driven world. Notwithstanding the emphasis on science and technology, the vision of Singaporean education has also been for the nation's young to discover their special talents and to develop those abilities to the fullest potential (Ministry of Education Singapore 2012a). Fostering a passion for lifelong learning in students is an important part of this process.

Science is compulsory for all students from Primary 3 onward through to lower secondary schooling (to Year 8). Students follow a general integrated science curriculum that is designed to introduce basic concepts about the natural and physical world. In primary science, five thematic learning modules (diversity, cycles, systems, energy, and interactions) help students make links between everyday life and what they learn in the classroom. Science teachers use thematic inquiry questions to "engage students in uncovering the important ideas at the heart of each theme" (Ministry of Education Singapore 2013a, 6). "Science as an inquiry" is a teaching approach that aims to equip students with the scientific knowledge, process skills, and ethics to help them understand "the roles played by science in daily life, society and the environment" (Ministry of Education Singapore 2013a, 1). Science content in lower secondary science is organized into four thematic modules—diversity, systems, interaction, and models (Ministry of Education Singapore 2006a). An additional module, "scientific endeavor," helps students deepen their understanding of the nature of science, the application of science and technology in daily life, and the social, environmental, and ethical issues that arise from the use of science and technology (Ministry of Education Singapore 2012c). In essence, Singapore's compulsory science learning program prepares students "to be sufficiently adept as effective citizens, able to function in and contribute to an increasingly technologically-driven world" (Ministry of Education Singapore 2013a, 1).

Students at the upper secondary (Years 9 and 10) and pre-university levels (Years 11 and 12) are offered science as an elective subject, study of which leads to GCE certification at the normal level (GCE N-level), the ordinary level (GCE O-level), or the advanced level (GCE A-level). Every student at the upper secondary level must take at least one science subject, which they can select from the various science options. Combined science options are also available (e.g., chemistry/physics, chemistry/biology, biology/physics), as are single "pure" science subject options—biology, chemistry, and physics. More academically inclined students can take up to three single science subjects. The subjects that upper secondary students choose and their subsequent performance on the national GCE N-level or O-level examinations determine their next education pathway, namely pre-university, polytechnic, or technical institute.

Students in the pre-university science track may read up to two single science subjects. The popular two-subject options are chemistry/physics and chemistry/biology. Chemistry, a prerequisite for future university medical, chemical, and bio-medical engineering programs, is a popular science subject. Less science-inclined students can take science subjects at a comparatively lower level involving less curriculum time. Students with exceptional aptitude and interest in science can pursue their science subjects within a special program where they learn more advanced topics and can also do research work under the supervision of university personnel.

Teacher Academic and Professional Education and Training

Singapore is committed to having suitably qualified science teachers for students at all levels of the nation's education system. Singapore's National Institute of Education (NIE), the sole teacher-training institute, works in close collaboration with the Ministry of Education and schools to train teachers for the national school system. Its three initial teacher preparation (ITP) programs target three groups of aspiring teachers (National Institute of Education Singapore n.d. a).

- The two-year Diploma in Education program prepares trainee teachers to become generalist teachers in primary schools.
- The four-year Bachelor of Science (Education) is an undergraduate program that provides students with both academic and professional development. Students enter one of two tracks. The primary track qualifies them to teach in primary schools, while the secondary track qualifies them to teach in secondary/pre-university schools.
- The one-year Postgraduate Diploma in Education (PGDE) program provides professional certification for university graduates. Aspiring teachers are carefully selected into one of three training tracks—primary, secondary, and junior college teaching.

In summary, each ITP program equips trainee teachers with the knowledge and skills they need to competently meet the demands and challenges of teaching a particular subject to a particular age group of students.

To be selected for the PGDE (Secondary) science-teaching track, candidates must have a science or engineering degree related to the major science subject they intend to teach. However, this track also prepares teachers to teach science as a "minor" subject, in which case candidates need to have obtained a good grade in the relevant subject at the GCE A-level or IB higher level, or to have studied the subject for at least one year at the university level. Those trainee teachers deemed to have inadequate content knowledge for science teaching as a minor subject have to take, in addition to the course requirements of the PGDE program, courses that provide them with the requisite knowledge.

Since 2009, trainee teachers intending to teach at the pre-university level have had to qualify through the PGDE (Junior College) teaching track. Stringent academic requirements are applied to select suitably qualified candidates. First and foremost, they must have at least an Honor's degree in the physical or biological sciences, usually in the science subject they want to teach. They must also have studied the particular science subject at the A-level or IB higher level. Candidates who are chemical engineering graduates with an Honor's degree can be accepted into the program that trains them to teach pre-university chemistry. Likewise, candidates who are electrical, mechanical, or civil engineering graduates can train to teach pre-university physics if they have at least a second class upper Honor's degree. However, candidates who have not undergone the pre-university course prior to their university experience (in Singapore, individuals holding a polytechnic diploma can study at university) cannot teach at the pre-university level.

Selected candidates undertake comprehensive and intensive initial teacher preparation. The one-year PGDE program has four main components—education studies, curriculum studies, teaching practice, and language enhancement and academic discourse skills training (LEADS). Education studies focuses on key educational concepts, principles, and aspects for reflective practice in schools. Curriculum studies requires trainees to study the methodology of teaching the specific curricular contents at primary, secondary, and junior college levels. Trainees working toward the PGDE (Secondary) specialize in the teaching methods associated with two subjects at the secondary school level, while candidates studying for the PGDE (Junior College) specialize in methods of teaching one subject at junior college level as well as one subject at secondary school level. The latter subject must align with the candidates' junior college teaching subject. Junior college PGDE candidates also engage in "knowledge skills training" to help them develop the values, knowledge, and skills expected of junior college teachers.

Practicums are compulsory for all trainee teachers. Each trainee has to do two stints of practice in schools. Before each practicum, the NIE, with input from the Ministry of Education, contacts the respective schools to arrange and ensure they can provide opportunities for the trainee teachers to teach their subjects of specialization. During the practicum, each trainee teacher is mentored by an NIE supervisor, who periodically visits him or her in school to observe and advise on his or her teaching. Each trainee is also supported by in-school mentors, the school's coordinating mentor, and cooperating teachers.

Trainee teachers furthermore participate in two special programs. The first is the Group Endeavours in-Service Learning (GESL), a community outreach service-learning program that provides trainees with opportunities to perform community services facilitated by an NIE facilitator. The second is the Meranti Project, a Ministry of Education-funded personal and professional development program that provides trainee teachers with opportunities to experience social-emotional learning, share personal aspirations with peers, and express their opinions in an open and creative environment.

Continuing Professional Development

The professional development of teachers does not stop at their initial teacher preparation. The Ministry of Education has invested in various professional development programs so that it can ensure every inservice teacher experiences continuing professional development (CPD) throughout his or her career. A key recommendation of the 1999 review of teacher training entitled every teacher to 100 training hours per year so that they could refresh their skills and knowledge in line with the latest understandings and developments in education (Ministry of Education Singapore 1999).

Leveraging on technology, the ministry put in place a centralized online training administration system, TRAISI. It advises teachers of professional development courses and also allows them to register for those courses. In 2001, the ministry also launched an e-learning system, the Virtual Institute of Training and Learning (VITAL), which offers teachers convenient, flexible, and timely learning through several online courses (Ministry of Education Singapore 2001). The ministry works closely with various service providers, such as NIE, to have on hand a wide variety of courses/programs designed to meet teachers' professional needs in the areas of skills and knowledge. More specifically, these courses help teachers upgrade their knowledge, keep abreast of new developments and initiatives in education, receive updates about pedagogical innovations, gain new competencies responsive to societal needs and demands, gain training in research and management skills, and enhance their teaching effectiveness through a commitment to lifelong learning.

Science teachers can participate in various inservice courses and workshops conducted by tertiary institutions and the Academy of Singapore Teachers. For example, science teachers interested in upgrading their subject content knowledge first identify a suitable course, say a 39-hour certification course titled "Environmental Health and Toxicology" conducted by NIE. They then ask their school's principal for approval to register in the course and, having received that permission, they register through the TRAISI system. Science teachers can avail themselves of other modes of learning in order to upgrade their content knowledge and pedagogical knowledge and skills. These include attendance at conferences, seminars, and talks, participation in mentoring programs, and collaboration in educational research.

Science Teacher Professional Associations

Singapore has two professional bodies dedicated to furthering the interests of science teachers and science education. The Science Teachers' Association of Singapore, founded by a group of enthusiastic science teachers and officially registered in 1965, promotes science education through courses, workshops, seminars, and field trips with the aim of benefiting not only science teachers but their students as well. For instance, in 2014 the association organized a lecture, as one of its Distinguished Science Lecture Series, during which science teachers listen to and interact with an eminent science educator. The association also organized two events for students: a "science camp," during which secondary school students lived, worked, and learned together on the theme "Innovative Sustainable Solutions for the 21st Century," and a drama competition promoting science learning for primary students. In addition, the association sponsors an annual award for excellence in science teaching.

Established in 2010, the Academy of Singapore Teachers aims to build a teacher-led culture of professional excellence for the teaching fraternity (Academy of Singapore Teachers 2012a). One special feature of the academy's work is its various network learning communities, grouped by teaching subject, teacher role, and interest. The objective of these communities is to nurture and support teachers in their teaching roles and practices. Teachers participate in subject groups called subject chapters for each of the three sciences of biology, chemistry, and physics. Each subject chapter is led by a core team of teachers recognized as among the best in their subjects, specialists from the Ministry of Education, and academics from NIE. The role of these leaders is to enhance teachers' curriculum knowledge, pedagogical skills, and assessment literacy.

Issues in Upper Secondary Science Teacher Quality

Teacher Supply

The teacher supply and staffing situation in Singaporean schools is currently favorable. In 2000, Singapore's teacher workforce was 24,000 strong. A decade later, it had grown to 30,000 because of the Ministry of Education having ramped up its teacher recruitment efforts to support its various initiatives (Ministry of Education Singapore 2010a). Improving teacher quality remains a priority as Singapore pursues its agenda of educational excellence. The teacher selection and training processes likewise remain rigorous. Aspiring teachers must have not only good academic credentials that meet the requirements of NIE's training programs but also a keen interest in teaching, strong personal attributes and values, and the ability to communicate and relate to the young (Ministry of Education Singapore 2014d).

At present, short-listed applicants are from the top one third of a graduating secondary school cohort (Centre on International Education Benchmarking 2014), and they go through a selection interview chaired by a panel of senior Ministry of Education officers. Those selected spend a compulsory six months to a year as untrained teachers in a school so that they can experience the realities and demands of teaching before being admitted into the NIE for professional training. This process allows both the ministry as employer and the prospective teacher to assess his or her qualities and abilities for a teaching career.

In addition to recruiting high-caliber applicants, the ministry recruits professionals from other fields who want a mid-career switch to teaching. These individuals also undergo stringent selection and training processes. In 2013, career-switch teachers formed 25 percent of the teaching workforce, up from 15 percent in 2002; in the last three years, 35 percent of recruited career-switch teachers have had at least three years of previous working experience (Ng 2014). The professional maturity that these teachers bring to the teaching fraternity in terms of their variety of knowledge, experience, and perspectives is highly valued. Engineers, for example, can breathe life into physics lessons when they tell students how they applied physics principles in their former jobs. Sharing former work experiences in this way can inspire students to take up science, engineering, or technology-related courses at the tertiary level and/or go into science-related careers.

By 2013, the Ministry of Education had achieved its target of 33,000 teachers (Ministry of Education Singapore 2014a). The ministry's teacher recruitment objective is currently that of ensuring there are sufficient teachers to replace teachers who retire or resign. Singapore presently has no shortage of qualified teachers, especially science teachers, at both the secondary and pre-university levels. In fact, some schools have had an excess of science teachers and have deployed them to co-teach classes where students need more help and attention or to personally mentor academically inclined students in their science projects. Alternatively, these teachers are encouraged to take professional development leave to pursue postgraduate studies.

Professional Education and Training Issues

The rigorous process of recruiting new teachers is but a first step toward teacher quality. Equally important is the process of professional education and training to equip new teachers with good foundational subject knowledge and/or pedagogical skills at the onset of their teaching career. At NIE, different initial training programs (ITPs) are available depending on new teachers' level of education at point of entry and, where appropriate, their choice of teaching subject and teaching level (primary, secondary, or junior college). After graduating from NIE's ITPs, all beginning teachers undergo a comprehensive induction program conducted by the Academy of Singapore

Teachers. This two-year program includes a three-day "Beginning Teachers' Orientation Programme" that aims to "consolidate their learning and emphasise the importance of nurturing the whole child" (Academy of Singapore Teachers 2012, para 6). It also includes inservice courses on classroom management, engaging parents, basic counseling skills, reflective practice, and assessment. Beginning teachers additionally participate in a "structured mentoring program" (Ministry of Education Singapore 2006b). Here, senior teachers, subject heads, or heads of department coach and guide these new teachers in various pedagogical tasks and understandings, such as teaching difficult concepts and skills, assessing and addressing student difficulties, handling off-task behaviors, performing administrative duties, learning the workings of the school, and meeting the expectations of school leaders.

The model of initial preparation for new teachers described in the preceding section has served the Singapore education system well over the past decade. Having met its target workforce of 33,000 teachers, Singapore has turned its attention to getting the best into teaching and the best from teachers. An issue currently being debated is whether a longer period of professional education and training would be better for developing new teachers and enhancing teacher quality. Many commentators deem the one-year intensive professional training program limited in terms of providing trainee teachers with the time they need to reflect on their learning and to relate the theories they have learned during their courses to the classroom and school practices they observe or experience during their practical training in schools. A current opinion is that a four-year undergraduate program might be the best option in terms of allowing trainee teachers to progressively develop their knowledge and skills, as well as their values, professionalism, and identity.

Issues at the Chalkface

Inquiry-based teaching and learning is currently the core of Singapore's school science curriculum, and orienting teachers to practice this way has been a challenge. Aimed at providing students with skills and processes that promote their critical and inventive thinking, inquiry-based science requires students to be active inquirers and teachers to be leaders of inquiry. The crux of the matter is teachers' competencies and confidence as facilitators of inquiry-based classrooms. Preservice training may equip them with the basic implementation know-how of inquiry-based lessons, but a lack of familiarity and/or experience with the demands of these lessons, such as scaffolding the inquiry process, managing group work, and facilitating collaborative learning among students, can hinder overall effectiveness.

Class size is another barrier to teachers' implementation of inquiry-based science. In Singapore, class sizes at upper secondary level are, as at most other secondary levels, about 40 students. This large class size may affect science teachers' teaching effectiveness, and especially so for a less experienced beginning teacher who has to struggle with classroom management skills while simultaneously coping with managing the teaching and learning processes.

Singapore assigns considerable importance to practical work in science learning. All national schools have well-equipped science laboratories manned by adequate numbers of laboratory technicians who assist science teachers prepare apparatus, instruments, and reagents for their practical lessons. Since 2006, school-based science practical assessments (SPAs) have replaced the traditional one-off practical examinations conducted as part of the GCE O-level and A-level science examinations. SPAs evaluate students' practical skills throughout the series of practical sessions in their science course or courses. Because this assessment constitutes 20 percent of an O-level pure science subject score and 15 percent of an A-level science subject score, teachers spend a significant amount of time and effort preparing students for the practical assessments. An observed but unintended backwash effect has been students' daily practical work being restricted to only a limited range of science experiments because of their teachers tending to set only experiments that are relevant for the O-level or A-level practical assessments.

Another issue focuses on the concern of how and why such curriculum planning and emphasis often runs contrary to the directions intended in actual classroom practice. If students do not know what they are doing and what they should be doing during practical work, then they probably do not understand the purpose of that work. For example, they may not know why certain procedures, apparatus, and reagents are required and/or how their practical work links to the theory that they learn. This situation potentially reduces practical work to "a mechanical level with little intellectual involvement" (Tan, Goh, and Chia 2001, 229). The Tan et al. (2001) study was conducted in Singapore and looked at students' learning of qualitative analysis during the practical work associated with their basic inorganic chemistry lessons. The study revealed that the concern expressed above was indeed happening in some of the schools that participated in the study. The authors surmised this outcome could be due to teachers not highlighting the purpose of and the theories underpinning the experiments and the procedures involved, or because teachers considered these aspects as not central to science experiments, assuming instead that the focus needed to be only on the data collected during practical work.

Another challenge with respect to teaching and learning science is the emphasis on high-stakes examinations in Singaporean education. As noted above, at the upper secondary and pre-university levels, the two national examinations—the GCE O-level at the end of Secondary 4 (Year 10) and the GCE A-level at the end of Year 2 junior college (Year 12)—are the milestone assessments, and teachers dedicate much effort and time to preparing their students for them. For instance, teachers feel compelled to complete the examination syllabus early so that they can have more time to go through past years' examination questions with their students. As a consequence, inquiry-based learning activities often take a back seat and do not feature to the extent set down in the science curriculum.

Next, there is the issue of workload. Singaporean teachers teach on average about 15 hours per week and spend twice as much time on teaching-related duties, such as preparing for lessons, providing remediation for weaker students, setting and marking students' homework, and conducting examinations (Ministry of Education Singapore 2013b). The heavy workload often leaves teachers with little time to review lessons in class, to self-reflect on their teaching, or to develop new materials for teaching. School leaders have the flexibility to deploy teachers in various roles beyond classroom teaching in order to ensure the school's program accords with the ethos in Singapore schools of helping every student learn and develop holistically. This expectation of efficient and high-quality work adds to the work demands on every Singaporean teacher.

In recent years, the Ministry of Education has invested in more human resources to support schools in developing their teaching and learning initiatives and innovations. As evidenced by the measure of pupil–teacher ratio (PTR), Singapore improved from a PTR of 19 in 2000 to 14 in 2012, which is comparable to the OECD's average (Ministry of Education Singapore 2013c). The ministry considers that an indicator additional to the PTR of teacher workload is needed in order to account for factors such as preparation for classes, marking of assignments, and guiding students who need help. As the ministry points out, a teacher's work does not consist only of face-to-face student teaching.

Yet another challenge science teachers face is the confusion many students feel when they are taught different models and theories developed to explain scientific phenomena. A case in point is the deeply entrenched Bohr model of an atom that students learn in secondary chemistry and which later causes dissonance with their learning of the new quantum mechanical model in pre-university chemistry. In a similar vein, Tan et al. (2005) found that students were using the octet rule framework, most likely carried forward from the learning of bonding in prior secondary science, to explain ionization energy trends in upper secondary chemistry. This type of issue may be the product of Singaporean teachers not explicitly teaching the nature of scientific models or emphasizing to students that they will learn different models at different stages of their education. For instance, students need to know that a model they learn at a particular stage may adequately explain a phenomenon at that level. However, as they progress to higher schooling levels, they need to be aware that they will learn the more advanced models in order to provide them with explanations of the same phenomenon at a deeper level.

Continuing Professional Development Issues

Since the mid-1990s, the Ministry of Education has introduced several educational reforms and initiatives. In 2002, it conducted a review of junior college and upper secondary education, followed by a review of primary education in 2009 and a review of secondary education in 2010 (Ministry of Education Singapore 2002b, 2009, 2010b). More recently, the ministry has reviewed polytechnic/vocational education (Ministry of Education Singapore 2014e). Two initiatives, "Thinking Schools, Learning Nation" and "Teach Less (to Enable Students to)

Learn More," were implemented in 1997 and 2004 respectively (Ministry of Education Singapore 1997b, 2004b), while three master plans focused on information technology (IT) and covering the periods 1997–2002, 2003–2008, and 2009–2013 were launched as part of Singapore's efforts to bring teaching and learning up to the standards required in today's twenty-first-century classroom environment (Ministry of Education Singapore 1997a, 2002a, 2008).

Singaporean teachers have been at the forefront driving these reforms and initiatives, and they have been urged to keep pace by continuously developing the pedagogical and technological knowledge and skills they need to facilitate student learning rather than to act as dispensers of knowledge. To help teachers help themselves to become lifelong learners acquiring current skills and knowledge, the Ministry of Education provides all teachers with 100 training hours per year; in some instances, teachers may be relieved of their regular workload to enable their participation (Ministry of Education Singapore 1999). Although some teachers initially responded lukewarmly to this provision, a culture of lifelong learning has been taking root within the teaching fraternity, especially in recent years. However, several pertinent issues remain with regard to continuing professional development (CPD).

An intractable issue is time. Teachers' heavy work demands and commitments in schools often prevent them from pursuing CPD while in normal service or using their 100 training hours per year entitlement. There is also the related issue of teachers' personal well-being and their need to have a good work/life balance. It is highly likely that the work productivity of teachers who are over-stretched and who have to add professional development and training to their normal workload is being compromised.

Another issue is how to cater for teachers' different professional development needs. A one-size-fits-all model for professional learning is unlikely to resolve this issue. Some teachers may need a timely refreshment of their science knowledge pertaining to a particular new teaching assignment; others may need specific training in new pedagogical approaches relating to similar purposes. Experienced teachers may benefit more from professional education on translating research into classroom practices. This diversity means that teachers' professional development needs to be conducted through courses, workshops, and programs that target different teacher needs and address specific pedagogical skills or content knowledge gaps, with these identified in a timely manner.

Trends and Developments in Upper Secondary Science Teacher Quality

Singaporean education has improved significantly over the last decade, and this achievement has attracted international recognition. Singapore's students have been ranked among the best performers in international assessments of student achievement in mathematics, science, and reading literacy. The three most prominent of these assessments are the Trends in International Mathematics and Science Study (TIMSS), the Programme for International Student

Assessment (PISA), and the Progress in International Reading Literacy Study (PIRLS), all of which are conducted every few years. Singaporean students have excelled in all of these assessments, with that performance being confirmed four times over in the TIMSS mathematics and science surveys (i.e., in 1999, 2003, 2007, and 2011) and twice in PISA (2009, 2012). The Singaporean students who took the 2012 PISA test were the highest ranked internationally with respect to their problem-solving skills.

These accolades reflect the high quality of Singapore's teaching workforce. Another large-scale cross-national survey, OECD's 2013 Teaching and Learning International Study (TALIS), affirmed the quality and commitment of Singapore's teachers (Ministry of Education Singapore 2014g). As highlighted in a report produced under the auspices of McKinsey & Company and titled "How the World's Most Improved School Systems Keep Getting Better" (Mourshed, Chiijoke, and Barber 2010), Singapore's education system actively intervenes to build up teachers' instructional skills and capabilities. It invests heavily in various professional development packages and milestone programs to ensure that every teacher receives CPD.

In 2007, Singapore implemented its first GROW package (Ministry of Education 2006c). This acronym refers to the development (growth) of education officers through better recognition, opportunities, and a focus on well-being. A year later, in 2008, the ministry followed up its first GROW package with an enhanced package titled "GROW 2.0" (Ministry of Education 2007). The aim of both packages was to build teacher and school leadership capacity in order to deliver reforms at the school level. Working in close collaboration with NIE and schools, the ministry remains intent on developing and implementing different measures directed toward encouraging teachers to develop professionally. For example, teachers can pursue teaching subject content or professional postgraduate studies under the Professional Development Continuing Model sponsorship scheme implemented in 2005 (Ministry of Education 2004a).

Directly relevant to science teachers' professional development are NIE's two Master's degree programs, the Master of Science (Life Sciences) and the Master of Education (Science). As mentioned earlier, science teachers can also participate in ad hoc inservice courses, seminars, workshops, and talks conducted by tertiary institutions, the Academy of Singapore Teachers, and the Science Teachers' Association of Singapore. Science teachers can furthermore avail themselves of multiple modes of learning, such as conferences, mentoring, research-based practice, network learning, and reflective practice. The ministry's launch of its Teacher Growth Model in May 2012 further consolidated efforts to encourage teachers to continually upgrade their knowledge and skills as well as take ownership of their professional growth and personal well-being (Ministry of Education Singapore 2012b).

At the school level, teachers have opportunities to share, collaborate, and co-develop new and better ways of teaching. For instance, science teachers are provided with "white space" within professional development communities (PLCs) to develop such areas as lesson plans and to refine instructional materials, as well as teaching strategies and assessment practices. The subject chapters (mentioned earlier in this chapter) support PLCs in schools by providing teaching resources and instructional leadership in their subjects. They also identify and share good pedagogic practices throughout the school system. To date, more than half of Singaporean schools have formed PLCs (Ministry of Education Singapore 2014f).

The NIE, a key leader in teacher professional education and training, works closely with the Ministry of Education and schools to provide high-quality relevant programs for teacher development. The institute recently redesigned its four-year undergraduate ITP program in order to attract Singapore's best young adults into teaching. It also launched its Teaching Scholars Programme (TSP) in 2014 (National Institute of Education Singapore n.d. c). This four-year program has added a multi-disciplinary curriculum to the core curricula of the Bachelor of Arts/Science (Education). Offering a wide range of electives, seminars, and leadership programs, the TSP aims not only to broaden students' perspectives both locally and globally but also to produce graduates with intellectual rigor and strong leadership skills. For Singapore, ITP programs such as the TSP are integral to raising teacher quality and in turn the caliber of teaching and learning, including, of course, science.

Recognizing the time constraints that teachers experience, the NIE in collaboration with the ministry has been customizing its courses and programs in an effort to address this difficulty. For instance, teachers wanting to study toward higher degrees can do so through a flexible part-time learning schedule under Ministry of Education sponsorship. Recently, the ministry put a pathway in place to encourage more teachers to participate in professional learning and development (National Institute of Education Singapore n.d. b). The special features of this pathway include modular learning so that teachers can take course modules at their own time and pace instead of trying to engage in a full program of study all at once. Teachers can thus accumulate over a longer period of time the academic credits that build toward higher certification.

The NIE also offers a variety of courses in modular form. For example, science teachers wanting to refresh their content knowledge can take short courses such as "Chemistry: Separation Techniques" or "Physics: Teaching Relativity." Other courses such as "The Design and Learning of 'Flip' Science Lessons" and "Active Learning Instruction in Physics with Data Loggers" provide teachers with direction on how to use computer technology during science teaching. Teachers can gain pedagogical knowledge and skills through courses such as "Inquiry Biology for Secondary Schools," "Inquiry Chemistry for Secondary Schools," and "Modelling Instruction in Chemistry," while teachers who want to engage in professional learning that leads to higher certification have at hand the Master of Education program with its different specializations. Of special relevance to science teachers are the specializations in science education, curriculum studies, and educational assessment.

As the authors of another McKinsey report have aptly pointed out, "an education system cannot exceed the quality of its teachers" (Barber and

Mourshed 2007, 16). Singapore is committed to preparing its young people for life in the twenty-first century and to ensuring that the city-state has the highly skilled workforce it requires for its high-value, technology-intensive manufacturing industries and world-class service sector (Ministry of Trade and Industry Singapore 2012). To meet this commitment, Singapore needs a high-quality, dedicated teaching workforce, among whom are high-quality science teachers able to provide the mainstay of educational excellence in Singapore's upper secondary science.

References

Academy of Singapore Teachers. n.d. *Teacher-Led Workshops 2014*. Singapore: Author, http://www.academyofsingaporeteachers.moe.gov.sg.

———. 2012. *Beginning Teachers' Induction Programme (BTIP)*. Singapore. Author, http://www.academyofsingaporeteachers.moe.gov.sg/professional-growth /professional-development-programmes/beginning-teachers-induction-programme -btip.

Barber, Michael, and Mona Mourshed. 2007. *How the World's Best-Performing Schools Come Out on Top*. New York: McKinsey & Company, http://mckinseyonsociety .com/ downloads/reports Education/Worlds_School_Systems_Final.pdf.

Center on International Education Benchmarking. 2014. *Singapore: Teacher and Principal Quality*. Washington, DC: Author, http://www.ncee.org/programs-affiliates /center-on-international-education-benchmarking/top-performing-countries /singapore-overview/singapore-teacher-and-principal-quality/.

Department of Statistics Singapore. 2014. *Population Trends 2014*. Singapore: Ministry of Trade and Industry.

Ministry of Education Singapore. 1997a. "Speech by Radm Teo Chee Hean, Minister for Education and 2nd Minister for Defence at the Launch of the Master Plan for IT in Education on Opening New Frontiers in Education with Information Technology" (press release, April 28, 1997, Suntec City, 10:00 a.m., Singapore), http://www .moe.gov.sg/ media/speeches/1997/280497.htm.

———. 1997b. "Speech by Prime Minister Goh Chok Tong at the Opening of the 7th International Conference on Thinking" (press release, June 2, 1997, 9:00 a.m., Suntec City Convention Centre Ballroom, Singapore), http://www.moe.gov.sg /media/speeches/1997/020697.htm.

———. 1999. "Dr Wong's Speech on Teacher Training" (speech March 17, 1999, Committee of Supply, Ministry of Education), http://www.moe.gov.sg/media /speeches/1999/sp170399a.htm.

———. 2001. "E-Learning @ MOE the Vital Way" (press release, July 13, 2001), http:// www.moe.gov.sg/media/press/2001/pr13072001.htm.

———. 2002a. "Speech by Mr Tharman Shanmugaratnam, Senior Minister of State for Trade and Industry & Education at Itopia 2002" (press release, July 24, 2002, 9:00 a.m., Suntec City, Singapore), http://www.moe.gov.sg/media/speeches/2002 /sp24072002.htm.

———. 2002b. "Government Accepts Recommendations for a Broader and More Flexible Curriculum and a More Diverse JC/Upper Secondary Education Landscape" (press release, October 15, 2002), http://www.moe.gov.sg/media/press/2002 /pr15102002.htm.

——. 2004a. "Speech by Mr Tharman Shanmugaratnam, Acting Minister for Education, at the NIE Teacher Investiture Ceremony" (press release, August 12, 2004, 2:30 p.m., Singapore Indoor Stadium), http://www.moe.gov.sg/media /speeches/2004/ sp20040812.htm.

——. 2004b. "Speech by Mr Tharman Shanmugaratnam, Minister for Education, at the MOE Work Plan Seminar 2004" (press release, September 29, 2004, 9:50 a.m., Ngee Ann Polytechnic Convention Centre, Singapore), http://www.moe.gov.sg /media/ speeches/2004/sp20040929.htm.

——. 2006a. "Over 40 Secondary Schools to Begin Offering Advanced Elective Modules" (press release, October 2, 2006), http://www.moe.gov.sg/media/press/2006 /pr20061002.htm.

——. 2006b. "Strengthening Teacher Development: The Structured Mentoring Programme for Beginning Teachers" (press release, January 26, 2006), http://www.moe .gov.sg/media/press/2006/pr20060126.htm.

——. 2006c. "MOE Unveils $250M Plan to Boost the Teaching Profession: New 'Grow' Package Strengthens Teacher Development and Recognition" (press release, September 4, 2006), http://www.moe.gov.sg/media/press/2006 /pr20060904.htm.

——. 2007. "Putting People at the Centre of the Education Enterprise: MOE Unveils 'GROW 2.0' Package to Further Strengthen Teacher Development and Recognition and Philosophy for Educational Leadership" (press release, December 28, 2007), http://www.moe.gov.sg/media/press/2007/pr20071228.htm.

——. 2008. "MOE Launches Third Masterplan for ICT in Education" (press release, August 5, 2008), http://www.moe.gov.sg/media/press/2008/08/moe-launches-third -masterplan.php.

——. 2009. Government Accepts Recommendations on Primary Education: Changes to Be Implemented Progressively Over the Next Few Years (press release, April 14, 2009), http://www.moe.gov.sg/media/press/2009/04/government-accepts -recommendat.php.

——. 2010a. "Speech by Dr Ng Eng Hen, Minister for Education and Second Minister for Defence, at the MOE Work Plan Seminar 2010" (speech September 23, 2010, 9:30 a.m., Ngee Ann Polytechnic Convention Centre, Singapore), http://www.moe .gov.sg/media/speeches/2010/09/23/work-plan-seminar-2010.php.

——. 2010b. "Strengthening Social-Emotional Support for Secondary School Students: Release of Secondary Education Review and Implementation (SERI) Committee's Report" (press release, December 28, 2010), http://www.moe.gov.sg /media/press/2010/12/strengthening-social-emotional-support-secondary-school -students.php.

——. 2012a. *Bringing out the Best in Every Child: Education in Singapore*. Singapore: Author, http://www.moe.gov.sg/about/files/moe-corporate-brochure.pdf.

——. 2012b. "New Model for Teachers' Professional Development Launched" (press release, May 31, 2012), http://www.moe.gov.sg/media/press/2012/05/new-model -for-teachers-profess.php.

——. 2012c. *Science Syllabus: Lower Secondary. Express/Normal (Academic)*. Singapore: Curriculum Planning & Development Division.

——. 2013a. *Science Syllabus: Primary 2014*. Singapore: Curriculum Planning & Development Division.

——. 2013b. "Parliamentary Replies: Teacher Workload and Resignation Rate" (press release, November 11, 2013), http://www.moe.gov.sg/media/parliamentary -replies/2013/11/Teacher%20workload%20and%20resignation%20rate.php.

———. 2013c. "Parliamentary Replies: Class Size, Student Outcomes and Teacher Work load" (press release, October 21, 2013), http://www.moe.gov.sg/media/ parliamentary -replies/2013/10/class-size-student-outcomes-and-teacher-workload.php.

———. 2014a. *Education Statistics Digest 2014*. Singapore: Author, http://www.moe .gov.sg/education/education-statistics-digest/files/esd-2014.pdf.

———. 2014b. *Nurturing Students*. Singapore: Author, http://www.moe.edu.sg/ education /nurturing-students/.

———. 2014c. *Our Education System*. Singapore: Author, http://www.moe.edu.sg /education/.

———. 2014d. *Teaching as a Career: Frequently Asked Questions*. Singapore: Author, http://www.moe.gov.sg/careers/teach/faqs/.

———. 2014e. "Speech by Mr Heng Swee Keat, Minister for Education, for Parliamentary Debate on the Applied Study in Polytechnics and ITE Review (ASPIRE) Report" (press release, September 9, 2014), http://www.moe.gov.sg/media /speeches/ 2014/09/09/speech-by-mr-heng-swee-keat-for-paliamentary-debate-on -the-applied-study-in-polytechnics-and-ite-review-report.php.

———. 2014f. "Keynote Address by Mr Heng Swee Keat, Minister for Education, at the Ministry of Education Work Plan Seminar 2014 (press release, September 23, 2014, 9.15 a.m., Ngee Ann Polytechnic Convention Centre, Singapore), http://www .moe. gov.sg/media/speeches/2014/09/23/keynote-address-by-mr-heng-swee-keat -at-the-ministry-of-education-work-plan-seminar-2014.php.

———. 2014g. "International OECD Study Shows a Quality, Dynamic and Committed Teaching Force in Singapore" (press release, June 25, 2014), http://www.moe .gov.sg/ media/press/2014/06/ international-oecd-study-shows-a-quality-dynamic -and-committed-teaching-force-in-singapore.php.

Ministry of Trade and Industry Singapore. 2012. *Manufacturing and Services*. Singapore: Author, http://www.mti.gov.sg/MTIInsights/Pages/Manufacturing-and -Services.aspx.

Mourshed, Mona, Chinezi Chijioke, and Michael Barber. 2010. *How the World's Most Improved School Systems Keep Getting Better*. New York: McKinsey & Company, http://www.mckinsey.com/client_service/social_sector/latest_thinking /worlds_most_improved_schools.

National Institute of Education Singapore. n.d.a. Initial Teacher Preparation Programmes. http://www.nie.edu.sg/study-nie/admissions/teacher-education-undergraduate -studies.

National Institute of Education Singapore. n.d. b. Professional Development Pathways. http://www.nie.edu.sg/study-nie/admissions/graduate-studies-professional -learning/professional-development-pathways.

National Institute of Education Singapore. n.d. c. NTU-NIE Teaching Scholars Programme. http://tsp.nie.edu.sg/index.html.

Ng, Jane. 2014. "More Switch Jobs to Become Teachers." *The Sunday Times*, February 9. http://www.straitstimes.com/breaking-news/singapore/story/more-switch-jobs -become-teachers-20140209.

Tan, Kim Chee Daniel, Ngoh-Khang Goh, and Lian-Sai Chia (2001). "Secondary Students' Perceptions about Learning Qualitative Analysis in Inorganic Chemistry." *Research in Science & Technological Education* 19 (2): 223–34.

Tan, Kim Chee Daniel, Keith S. Taber, Ngoh-Khang Goh, and Lian-Sai Chia. 2005. "The Ionisation Energy Diagnostic Instrument: A Two Tier Multiple-Choice Instrument to Determine High School Students' Understanding of Ionisation Energy." *Chemistry Education Research and Practice* 6 (4): 180–97.

Further Reading

Koh, Tommy, Lee Lin Chang, and Joanna Koh, eds. 2014. *The Little Red Dot: Reflections of Foreign Ambassadors on Singapore*, vol. 3. Singapore: World Scientific Publishing Company.

Lee, Kuan Yew. 2000. *From Third World to First: The Singapore Story, 1965–2000: Memoirs of Lee Kuan Yew*. Singapore: Times Editions, Singapore Press Holdings.

Saravanan, G. 2012. *Education and the Nation State: The Selected Works of S. Gopinathan*. London and New York: Routledge.

4

Canada

Todd M. Milford and Christine D. Tippett

Teaching Science in Canada

Backdrop

Canada has a total landmass of approximately 10 million square kilometers and is divided into 10 provinces (British Columbia, Alberta, Saskatchewan, Manitoba, Ontario, Quebec, New Brunswick, Nova Scotia, Prince Edward Island, and Newfoundland and Labrador) and three territories (Yukon, Northwest Territories, and Nunavut). Although Canada is the second largest country in the world, it ranks 37th by population, with 35.5 million people. Canada has two official languages, French and English. It is one of the most ethnically diverse nations in the world and has one of the highest per capita rates of immigration.

In Canada, education is a provincial or territorial responsibility, which means that instead of a national education system, there are 13 distinct, yet related, systems. Provincial and territorial departments or ministries of education are responsible for overseeing the organization, delivery, and assessment of education at the elementary, secondary, and post-secondary levels.

Most jurisdictions in Canada offer an education that spans from kindergarten (age 5) to Grade 12 (age 17/18), with the mandatory attendance age ranging from 5 to 7 for entrance and 16 to 18 for exit. School grade configurations may vary within a province or even within a school board. Common grade configurations are elementary schools (K–5, K–6, or K–8), middle schools (Grades 6–8, 7–8, or 6–9), junior secondary schools (Grades 8–9 or 8–10), or secondary schools (Grades 7–12, 8–12, or 9–12). A small proportion of secondary schools are configured as senior secondary (Grades 10–12 or Grades 11–12). The five less-populated provinces (Alberta, Newfoundland and Labrador, Nova Scotia, Prince Edward Island, and Saskatchewan) recently included Grades 10–12 as a typical configuration.

Although sports-related or performing arts specialty schools (e.g., hockey or music) are relatively common in Canada, science specialty schools are a rarity. Secondary school diplomas are awarded to students who complete a requisite number of compulsory and optional courses, as determined by particular provinces or territories. Canada's secondary school graduation rate was 85 percent in 2011 (Statistics Canada 2014).

Teachers in Canada are unionized as part of provincial and territorial collective agreements. Typically under such agreements, teachers secure teaching positions based upon seniority accrued while teaching within school districts. Teachers may possess postgraduate certification and/or additional degrees. While these subsequent qualifications equate with higher rates of pay, they do not necessarily mean better possibilities for job selection or secure employment and can actually make obtaining an initial teaching position more challenging.

School Science

The current aim of science education in Canada is built on the characteristics of a scientifically and technologically literate population. According to the Ontario Ministry of Education, "A scientifically and technologically literate person is one who can read and understand common media reports about science and technology, critically evaluate the information presented, and confidently engage in discussions and decision-making activities that involve science and technology" (2007, 3).

Science is part of the prescribed curriculum in all provinces and territories in Canada and is a required subject from kindergarten to Grade 10 inclusive. A range of elective science courses is offered to students in Grades 11 and 12. Although Canada does not have a standard science curriculum, the Common Framework of Science Learning Outcomes K to 12 (Council of Ministers of Education, Canada 1997) sets out a national vision for scientific literacy. The framework also lists foundation statements, outlines general and specific learning outcomes, and provides illustrative examples for some of these outcomes. It furthermore describes four goal areas of science literacy: science, technology, society, and the environment; skills; knowledge; and attitudes. The following descriptions are excerpted from British Columbia:

1. *Science, technology, society, and the environment (STSE).* Students will develop an understanding of the nature of science and technology, of the relationships between science and technology, and of the social and environmental contexts of science and technology.
2. *Skills.* Students will develop the skills required for scientific and technological inquiry, for solving problems, for communicating scientific ideas and results, for working collaboratively, and for making informed decisions.
3. *Knowledge.* Students will construct knowledge and understandings of concepts in life science, physical science, and Earth and space science,

and apply these understandings to interpret, integrate, and extend their knowledge.

4. *Attitudes.* Students will be encouraged to develop attitudes that support the responsible acquisition and application of scientific and technological knowledge to the mutual benefit of self, society, and the environment. (British Columbia Ministry of Education 2007)

The framework provided a common starting point for the development of science education within each participating jurisdiction, which is clearly evident from an analysis of provincial curriculum documents. For example, students in Grade 4 study "rocks and minerals" in all jurisdictions except British Columbia, Alberta, and the Yukon; students in these three latter jurisdictions study this topic at other grade levels. In Grades 9 and 10, the science curriculum is general, with students' learning focused on the strands of Earth, life, physical, and chemical sciences, presented within the context of a single course. The influence of the framework is evident in the content of these courses. For example, all Grade 10 students in Canada study "chemical reactions." In some provinces, students can choose between academic (theory and abstract problems) and applied science courses (more practical, with applied applications).

In Grades 11 and 12, science courses typically emphasize the traditional domains of biology, chemistry, physics, and Earth and space science, and are designed to help students build a solid conceptual and procedural foundation in science. Students choose between course type (academic or applied) and content (domain) on the basis of their interests, achievement, and post-secondary goals. Graduation requirements typically include the successful completion of one Grade 11 science course, but there are many more optional science courses available for students in upper secondary school. Course selection at this level determines the pathways students can follow in subsequent grades as they prepare for university, college, or the workplace.

In 2012, just over two million students were attending universities and colleges in Canada. In total, just over 20 percent of these students were enrolled in courses in science and/or science-related areas. A recent report on science in Canada (Amgen Canada 2013) argued that students do not receive enough encouragement to take advanced science and mathematics courses. The report noted that less than 50 percent of students graduate with senior STEM courses, despite about 70 percent of Canada's top jobs requiring such preparation. Bordt et al. (2001) reported that 59 percent of upper secondary students stop taking mathematics and science because they consider these subjects difficult or boring. Interestingly, the students who said this included students who had done well in these subjects and who said they thought such classes were important to their futures. Currently, only two provinces (New Brunswick and Newfoundland) require a Grade 12 science credit for graduation. This lack of required upper-level courses means students may be making course selections that greatly limit their post-secondary pathways and career choices.

Canada has not experienced the same level of high-stakes testing and subsequent impact on teachers that has been, and continues to be, common within the American education system. That said, Canadian students are not excluded from testing, as they are assessed in science at the provincial, national, and international levels. Provincially, there is variability in how jurisdictions implement assessments, but typically some form of assessment in upper secondary science is required for graduation. For example, students in Nova Scotia enrolled in biology 12, chemistry 12, and/or physics 12 must sit the Nova Scotia Examinations, which contribute up to 30 percent of students' final course marks, while Alberta students enrolled in biology, chemistry, physics, and science at the upper secondary level are required to write diploma examinations that are worth up to 50 percent of their final course mark.

At the national level, the Pan-Canadian Assessment Program (PCAP), a measurement of the ability of students to use their learning skills to solve real-life situations in the areas of reading, mathematics, and science, is administered every three years. In 2013, the PCAP focused on science literacy (three competencies of science inquiry, problem solving, and scientific reasoning), four sub-domains (nature of science, life science, physical science, and Earth science), and attitudes about science and its role in society (Council of Ministers of Education, Canada 2014). In 2013, approximately 32,600 students and 1,500 schools from all provinces in Canada participated in the assessment, with 91 percent of students achieving at or above the expected level of performance in science.

Canada also participates in international assessments such as the Trends in International Mathematics and Science Study (TIMSS), which assesses students at Grades 4 and 8, and the Programme for International Student Assessment (PISA), which assesses students at age 15. TIMSS results show that the science achievement of Canadian students has been declining over the past decade. However, according to the most recent PISA results (i.e., 2012), Canadian 15-year-old students are among the best in the world in science. Among the 65 countries that participated in PISA 2012, only 7 outperformed Canada in science. However, although Canada continues to perform well in science, its international standing among PISA participants has slipped, and its average achievement score in science decreased significantly from 2006 to 2012 (Council of Ministers of Education, Canada 2013). Fazio and Karrow (2013) suggest that a potential reason for this trend is the emphasis on literacy and numeracy in governmental initiatives, which include provincial assessments such as the Foundation Skills Assessment tests administered to students in Grades 4 and 7 in British Columbia and the Education Quality and Accountability Office tests administered to students in Grades 3, 6, and 10 in Ontario.

Teacher Academic and Professional Education and Training

Teacher education program options in Canada are relatively varied. Currently, 450 different programs are offered across the country. These range from eight-month post-degree programs to six-year concurrent degree programs (Olson et al. 2015). Most often, admission requirements for Canadian preservice teacher education programs that lead to a Bachelor of Education (BEd) degree include both academic and non-academic components (e.g., work experience). These requirements have been described as higher than the standards for many other countries (Crocker and Dibbon 2008). Academic requirements include minimum marks that range from 65 percent to 75 percent (C+ to B+) and may also include the completion of specific courses.

BEd programs tend to be specialized by instructional level. Prospective education students must decide if they wish to teach in early childhood, elementary, primary/junior, middle school, junior/intermediate, intermediate/secondary, secondary, or adult education. Teachers at the elementary and primary/junior levels in Canada are usually generalists whose knowledge and skills span a range of subjects, including mathematics, English, science, art, health, physical education, and social studies. Middle school and junior/intermediate teachers may either be generalists or may be responsible for teaching several but not all subject areas. However, in secondary schools, teachers are more often specialists, frequently in two subject areas (e.g., physics and mathematics or chemistry and biology). In Canada, teacher education programs typically do not distinguish between upper and lower secondary. Admission requirements are no more stringent for teachers who want to teach at senior secondary schools than they are for other areas of specialization.

A recent examination of teacher education programs in Canada (Olson et al. 2015) revealed three general pathways for individuals who wish to become certified to teach secondary school science: direct entry from secondary school, consecutive entry after an initial Bachelor's degree other than a BEd, and concurrent entry. The latter sees students entering the program directly from secondary school and then completing the BEd and a subject degree concurrently. A typical example of each pathway to secondary science teacher certification follows.

- *Direct Pathway:* University of Regina (Saskatchewan)
  The University of Regina's four-year secondary education program consists of 120 credit hours and requires a major and a minor. Acceptable majors include biology, business, technology and media, chemistry, English, French, health, mathematics, physical education, physics, and social studies. Minors include arts, business, dance, drama, English, French, health, mathematics, music, outdoor education, physical education, religious education, science, social studies, special education, and visual education (University of Regina Faculty of Education 2013). The program also includes a 16-week practicum.

- *Consecutive Pathway:* Trent University (Ontario)
  Trent University's consecutive secondary education program requires students to complete an Honors Bachelor's degree with an emphasis in teacher education. Students need to complete five credits in their first teaching area and three credits in their second teaching area. The teaching areas recognized by the Ontario College of Teachers are biology, chemistry, computer studies, dramatic arts, English, French (second

language), geography, health and physical education, history, native studies, mathematics, physics, and visual arts. On completing this initial degree, students apply for the consecutive program, which consists of an additional academic year in a teacher education program and includes 17 weeks of practicum placement (Trent University School of Education 2014).

- *Concurrent Pathway*: University of Lethbridge (Alberta)
  The University of Lethbridge's concurrent degree is a five-year program that results in the simultaneous granting of two degrees (e.g., Bachelor of Science and Bachelor of Education). In addition to the requirements for the Bachelor of Science, the degree includes 20 courses taken through the faculty of education. Field experience consists of 27 weeks off campus in a broad range of school settings (University of Lethbridge Faculty of Education 2014).

Graduating from a teacher education program in Canada does not mean immediate entry to the profession, as teachers also need a license from a jurisdictional governing body, such as the Teacher Regulation Branch in British Columbia. This certification process typically includes a recent criminal record check, birth certificate and proof of citizenship, and education program transcripts. On graduating from a teacher education program in Canada and with certification, teachers may, depending on the jurisdiction, teach only the subjects for which they trained or they might teach across all content areas because of a shortage of teachers with training and experience in certain subject areas.

Continuing Professional Development

Professional development in Canada is viewed as "the continuous growth of personal and professional knowledge and expertise that enhances teaching in support of student learning" (Canadian Teachers' Federation 2004). The decentralized nature of the Canadian education system means that professional development is as varied as the pathways to secondary science certification. Professional development opportunities, including those for science teachers, are provided at the school, district, and provincial levels. For example, in British Columbia, formal professional development occurs over five non-instructional days during the school year.

Some jurisdictions offer formalized professional development comprised of accredited inservice programs. One example is the system of Additional Basic Qualifications (ABQs) and Additional Qualifications (AQs). These two sets of qualification are legislated and accredited by the Ontario College of Teachers and recorded on the teacher's Certificate of Qualification and Registration. Teachers who wish to pursue an in-depth investigation into a particular subject can take a three-part program that consists of skills and knowledge development, an expansion of skills and knowledge development, and development of leadership skills within the discipline.

Education experts in Canada have recently noted a shift from professional development and training to professional learning (Hurley 2015). The definition of professional development for teachers has been expanding to include a process of personal growth through programs, services, and activities designed to enable members, individually or collectively, to enhance professional practice.

Other, more informal, opportunities arising as a part of non-instructional days within a local school are also common. Unfortunately, there are few documented cases of secondary science professional development opportunities in Canada. However, a relevant example of a long-term, community-based professional development program in science can be given. The program was implemented in British Columbia and involved participating science teachers attending a number of workshops on topics related to enhancing science literacy through explicit language arts instruction. During its first year, the program adopted a drop-in format in order to encourage higher participation. In the following years, teachers who self-identified as teacher leaders during earlier sessions could access a series of five or more connected workshops. After five years, these teacher leaders collaborated with the researchers to create an instructional resource and to facilitate professional development workshops for their colleagues (Anthony, Tippett, and Yore 2010).

Science Teacher Professional Associations

Opportunities for professional development for upper secondary science teachers are also provided through provincial specialist organizations. Most provinces within Canada have associations for science teachers that host conferences, provide awards for teaching and professional development, and publish professional journals. For example, the Saskatchewan Science Teachers Society sponsors the biannual conference Sciematics, publishes the journal *Accelerator*, and provides ASSIST, a summer professional development program that includes activities, labs, and teaching strategies and is intended to help teachers learn more about science. The Alberta Teachers Association, with a membership of approximately 600 teachers (60 percent senior secondary, 35 percent junior secondary, and five percent elementary), sponsors the annual ATA Science Council Conference and publishes the *Alberta Science Education Journal*, which is refereed and issued at least twice a year. Although Canada does not have a national science teacher association, the National Science Teachers Association in the United States does have Canadian members.

Issues in Upper Secondary Science Teacher Quality

Teacher Supply

Canada generally does not experience difficulty attracting applicants to the teaching profession, which is usually held in positive regard. In addition, salaries for teachers are competitive with those of other Canadian public servants (Gambhir et al. 2008): "Canada is consistently able to recruit high-quality students into teaching, with the majority of prospective teachers drawn from the top 30% of their college cohorts" (Center for International Education Benchmarking 2015, para. 2). In many jurisdictions (e.g., Ontario), more people apply for teacher education programs than there are places. There is even some indication of an oversupply of prospective teachers graduating from teacher education programs. However, those individuals who are qualified to teach science in secondary school are currently in demand.

A 2010 study by the Ontario College of Teachers showed that two-thirds of education graduates from Ontario's class of 2009 reported themselves as unemployed or underemployed. However, this oversupply was not seen as equal across content areas or geographic regions, given the recruitment difficulties being experienced for subjects such as mathematics, physics, technology, and languages) and by schools in rural areas (Council of Ministers of Education, Canada 2008). Provinces including Alberta, British Columbia, and Nova Scotia reported teacher shortages in such areas as technology education, industrial education, human ecology, and senior secondary school science. Reasons suggested for these shortages included a lack of teachers with the desired specializations (e.g., chemistry and physics) and the higher salaries that individuals with science and STEM-related qualifications can obtain in other employment sectors, such as industry (Crocker and Dibbon 2008). These shortages mean that schools either do not offer upper secondary science courses or their teachers must cope with trying to teach content areas that fall outside of their preservice training and experience.

Professional Education and Training Issues

An issue related to the preparation of science teachers in Canada is the widely reported disconnect between the theory presented during teacher education programs and the practice preservice teachers recall from their own school experiences and see during their practicum placements (Crocker and Dibbon 2008; Gambhir et al. 2008). This disconnect has also been described as "the divide between learning experiences in university-based coursework and those in school-based field experiences" (Falkenberg 2010, 4), a view that emphasizes differences between theory and practice rather than an appraisal of the quality of these experiences. According to MacDonald (2010, 261): "This divide may cause a lack of uptake of educational theory in the practice of teaching as well as the lack of practice-based influence on university-taught educational theory."

Preservice teachers may not easily adopt research-informed practices presented in methods courses. Because they have experienced pre-reform education themselves, they anticipate learning about pre-reform methods, so when faced with reform-based pedagogy, they can experience cognitive dissonance (Bencze, Hewitt, and Pedretti 2009). Critical to a consideration of science teacher education is the tension between research-driven theory of an inquiry-based, hands-on science curriculum and the frequent reality of "cookbook" laboratory exercises accompanied by textbook readings (Banilower et al. 2010).

In Canada, a growing awareness of the theory–practice divide has led to recent teacher education reforms aimed at strengthening the relationships between universities and the schools in which preservice teachers are placed (Crocker and Dibbon 2008). Currently, there are a number of ongoing efforts to help bridge the divide between preservice teachers' experiences in their teacher education programs and the realties they face once they move to the classroom (either as student teachers or after certification). Here we describe two such programs that have been documented in the literature.

Case 1: Teaching Science Methods Courses Off Campus

In an attempt to foster stronger links between preservice and inservice settings (the university and the high school), MacDonald (2010) taught a secondary science methods course off campus, in a neighboring high school. The course, which was separate from formal practicum placements, ran for two semesters. Its after-school schedule provided opportunities for interactions with the inservice teachers. Situating the course in a school exposed the preservice teachers to the pragmatic constraints of teaching, such as class sizes, equipment availability, and limits on photocopying. The off-campus setting also facilitated preservice teachers' efforts to connect pedagogical theory with the realities of practice.

Although the logistics involved in scheduling a methods course off campus can be challenging, MacDonald (2010) noted that both preservice teachers and inservice teachers valued the opportunity to learn from one another. The preservice teachers reported their willingness to adopt inquiry-based activities rather than more traditional direct instruction approaches, while the inservice teachers were invigorated by the presence of the preservice teachers. According to Gambhir et al. (2008), forging connections between theory and practice and preservice and inservice teachers helps address the reported dissonance between university teaching and school experience.

Case 2: Methods in Science and Technology (Engineering) Teacher Education

Case study approaches have been used to help preservice teachers envision the enactment of pedagogical theories presented in teacher education courses in actual classroom contexts. However, as Bencze et al. (2009) suggest, if preservice teachers do not see themselves in the case study, the connections between theory and practice may not be strongly developed. Seeking to make the case study approach more relevant to preservice teachers, and to more effectively bridge the theory–practice divide, Bencze and colleagues paired a video case study with subsequent teaching opportunities. The teaching opportunities were designed to foster deeper engagement with and critical reflection upon the pedagogical aspects highlighted in the video.

At the start of the secondary science and technology methods course, preservice teachers reported an almost total lack of experience with technological design (engineering) and a similar lack of interest in incorporating technological design with students. After participating in the video case study and teaching activities, many of the preservice teachers felt confident in their knowledge of technological design and suggested they would include it in future teaching. This case-study teaching/activity-reflection approach may help science teacher educators address preservice teachers' lack of confidence in teaching science and engineering.

Issues at the Chalkface

Despite the clearly articulated vision of the Common Framework, secondary science instruction in Canada still seems focused on content, a situation that reflects earlier visions of science literacy (Aikenhead 2007). The pervasiveness of more traditional approaches to science education is a contributory factor to the dissonance preservice science teachers experience when they learn about current ideals in their methods courses and then witness previous versions of best practices of science instruction during practicum placements (Bencze et al. 2009; Falkenberg 2010). Despite the best efforts of science teacher educators, novice secondary science teachers may enter the classroom with a limited conception of what it means to teach for the demands of twenty-first-century society (MacDonald 2010).

Current visions of science literacy incorporating twenty-first-century skills include hands-on inquiry approaches, authentic assessment, and student-centered pedagogy. These visions also position the teacher as facilitator, consist of fundamental and derived aspects, and emphasize the essential nature of language in doing and learning science (Aikenhead 2007; Norris and Phillips 2003; Tippett 2011).

Although the primary role of the Trends in International Mathematics and Science Study (TIMSS) is to report on international mathematics and science achievement trends for students at the fourth and eighth grades, it also provides ancillary data on selected factors that affect learning (Martin et al. 2012). For example, TIMSS collects information on the availability of science-based school resources such as books and information and communication technologies (ICT), the presence or otherwise of which can influence the quality of classroom instruction and student learning. In Canada, only the provinces of Quebec, Ontario, and Alberta participated in TIMSS 2011, providing a limited data set. Despite this limitation and the lower grade levels targeted by TIMSS, these data supply the most accurate picture of science teaching and learning conditions in Canadian schools.

According to the most recent iteration of TIMSS (i.e., 2012), the instruction of students in Grade 8 was affected by a shortage in science resources approximately 35 percent of the time in Quebec, 28 percent of the time in Ontario, and 38 percent of the time in Alberta. These results were far below the reported international average of 78 percent of the time. Quebec, Ontario, and Alberta reported that vacancies for Grade 8 science teachers were somewhat or very difficult to fill approximately 26 percent, 8 percent, and 8 percent respectively, compared with the international average of 19 percent. Schools were also asked about resource availability for conducting science activities. Quebec reported that 99 percent of schools had science laboratory facilities and 93 percent of teachers had assistance available to conduct experiments. Ontario reported 52 percent for the former and 13 percent for the latter, while Alberta reported 85 percent and 23 percent respectively, compared with the international averages of 80 percent and 57 percent.

Continuing Professional Development Issues

An extensive search of the literature reveals a lack of documented research (or publications in general) on secondary science teaching and secondary science professional development in Canada. A search of both Canadian journals related to science education (i.e., *Alberta Science Education Journal* and *Canadian Journal of Science, Mathematics and Technology Education*) as well as an international journal (*Professional Development in Education*) failed to locate a single article on professional development in upper secondary science in Canada. This lack of information suggests that Canada is presently experiencing little in the way of continuing professional development issues at this time.

The apparent lack of issues could be a product of high admission standards for teacher education programs, fairly high levels of regard for the teaching profession across the country, a greater supply of newly certified teachers than there are positions available (although supply and demand varies with location and subject), and continued above-average results on international measures such as TIMSS and PISA for Canadian students.

Trends and Developments in Upper Secondary Science Teacher Quality

The Teaching and Learning International Survey (TALIS), conducted by the OECD, is a large-scale international survey of the working conditions of teachers and learning environments (OECD 2013b). TALIS focuses on lower secondary school teachers (ISCED Level 2 as defined by UNESCO) and includes information on initial training, professional development, learning environments, and classroom climate. Although Canada as a whole did not participate in the two iterations of TALIS to date, the province of Alberta took part in 2013. The results, despite being limited to this single province (and to 1,773 teachers at 182 schools) and restricted to a lower secondary context, are presently the best source of information about secondary science teaching in Canada; currently, no other extensive data sets on teachers and teaching across the country exist. Although TALIS did not differentiate between subject areas, it is likely that the Alberta science teachers who participated in the survey responded similarly to Alberta teachers as a whole.

According to the TALIS 2013 results, the average lower secondary teacher in Alberta is female (60.3 percent), middle-aged (40.1 years old), and working full time (91.1 percent). This average teacher is based at a public school (95.1 percent), located in a rural area with fewer than 15,000 people (40.3 percent), has 12.9 years of teaching experience, and has completed a teacher education program (98.3 percent).

Of particular relevance to this chapter, the 29.4 percent of Alberta respondents currently teaching science have varying levels of education and training or professional development in science: 81.4 percent reported having ISCED Level 4 or above education and training (compared to an international average of 89.0 percent), while 10.1 percent of respondents reported no education, training, or professional development (compared to an international average of 7.6 percent). Additionally, the majority of the teachers in Alberta reported feeling "not at all prepared" or "only somewhat prepared" to teach the content of the subjects they teach (11.5 percent compared to the international average of 6.8 percent).

This feeling of ill-preparedness may be partially responsible for the popularity of science associations across Canada, although few Alberta respondents reported a need for professional development related to subject matter or pedagogy (2.6 percent and 2.4 percent respectively). The areas in which most Albertan teachers reported a need for professional development were new technologies in the workplace and ICT skills for teaching (11.8 percent and 9.3 percent respectively).

Today's students live in an ICT-rich world: "ICT is pervasive in social and economic life with Canada being one of the most Internet connected countries and having a population that rapidly appropriates new technologies for personal use" (Milton 2003, 2). Government funding for education often contains ICT-specific designations. For example, the Ontario Ministry of Education recently announced $150 million in funding for technology (Reiti 2014). Students therefore expect to see and to use ICT in their science classrooms.

ICT does offer a range of affordances for science teachers, such as supplementing hands-on investigations, providing access to information, enhancing motivation, and supporting pedagogy (British Educational Communications and Technology Agency 2003). The use of ICT in the upper secondary science classroom is closely tied to the development of twenty-first-century skills such as critical thinking, problem solving, creativity, innovation, social responsibility, leadership, collaboration, and communication (OECD 2013a; Partnership for 21st Century Skills 2009). These twenty-first-century skills are necessary so that individuals can fully function in the workforce, modern society, and the global economy (OECD 2013a). However, while digital technologies, animations, and texts may be readily available, changing pedagogy so that it emphasizes ICT use can be challenging. As the British Educational Communications and Technology Agency (2003) found, the requirement for science teachers to be proficient in the use of ICT effectively places considerable demands on them.

In addition to the new interpersonal and pedagogical skills that teachers must develop in order to effectively integrate ICT in their science classrooms, other factors that can pose challenges include teachers' lack of the following: confidence, experience, and training in using specific ICTs, timetabled access to ICT resources such as a school computer lab, dependability of ICT equipment and infrastructure (e.g., Internet connections), and adequate technical support at the school level (Milton 2003; Osborne and Hennessy 2003). Teachers must balance any desire to use ICT to facilitate science learning with the necessity to learn and teach ICT technical skills.

Canada identifies science as an important component of the education system, making it a universal requirement for all students up to Grade 11. Canadian students do well in science assessments at the national level (e.g., PCAP) and international level (e.g., PISA and TIMSS). However, several areas of concern are becoming evident. Although Canadian students' average scores on these assessments remain above the international averages, Canada's international rankings have been dropping, a trend that has been linked to the aforementioned increasing emphasis on literacy and numeracy in government initiatives (Fazio and Karrow 2013). Also, lower secondary teachers in Alberta report below the international average for levels of training and professional development as well as preparedness to teach science (OECD 2013b). This lack of preparedness may manifest as science experiences that negatively influence students' decisions to pursue science at the upper secondary and post-secondary level. Current student enrolment in upper secondary science courses in Canada is lower than is necessary to meet anticipated demands for STEM-related jobs, not to mention meeting the need for a more scientifically literate population. These issues in science teaching could be partially addressed by placing science on an equal footing with literacy and numeracy, improving science teacher education programs, providing additional opportunities for professional development in science teaching (particularly with respect to using ICT), and developing more stringent science requirements for high school graduation.

References

Aikenhead, Glen. 2007. "Expanding the Research Agenda for Scientific Literacy." Paper presented at the Linnaeus tercentenary symposium on "Promoting Scientific Literacy: Science Education Research in Transaction," Uppsala University, Uppsala, Sweden, 28–29 May 2007, https://www.usask.ca/education/profiles/aikenhead/webpage/expand-sl-res- agenda.pdf.

Amgen Canada. 2013. *Spotlight on Science Learning: The High Cost of Dropping Science and Math.* Mississauga, ON: Author, http://www.letstalkscience.ca/images/Research/Spotlight /LTS_Spotlight2013_Executive_Summary.pdf.

Anthony, Robert J., Christine D. Tippett, and Larry D. Yore. 2010. "Pacific CRYSTAL Project: Explicit Literacy Instruction Embedded in Middle School Science Classrooms." *Research in Science Education,* 40 (1): 45–64. doi:10.1007/s11165-009-9156-7

Banilower, Eric, Kim Cohen, Joan Pasley, and Iris Weiss. 2010. *Effective Science Instruction: What Does Research Tell Us?* 2nd ed. Portsmouth, NH: Center on Instruction AT RMC Research Corporation, www.centeroninstruction.org.

Bencze, Larry, Jim Hewitt, and Erminia Pedretti. 2009. "Personalizing and Contextualizing Multimedia Case Methods in University-Based Teacher Education: An Important Modification for Promoting Technological Design in School Science." *Research in Science Education*, 39 (1): 93–109. doi:10.1007/s11165-007-9076-3

Bordt, Michael, Patrice de Broucker, Cathy Read, Shelley Harris, and Yanhong Zhang. 2001. "Determinants of Science and Technology Skills: Overview of the Study." *Education Quarterly Review, Statistics Canada*, 8 (1): 8–11, http://publications.gc.ca/Collection-R/Statcan/81-003-XIE/0010181-003-XIE.pdf.

British Columbia Ministry of Education (BCME). 2007. *Science K to 7: Integrated Resource Package 2005*. Vancouver, BC: Author, https://www.bced.gov.bc.ca/irp/pdfs/sciences/2005scik7_4.pdf.

British Educational Communications and Technology Agency (Becta). 2003. *What the Research Says About Using ICT in Science*. Coventry: Author, http://www.mmiweb.org.uk/publications/ict/Research_Motivation.pdf.

Canadian Teachers' Federation. 2004. *Policy on Professional Development*. Ottawa, ON: Author, http://www.ctf-fce.ca/Documents/PDPolicy-English.pdf.

Center for International Education Benchmarking. 2015. *Canada*. Washington, DC: Author, http://www.ncee.org/programs-affiliates/center-on-international-education-benchmarking/top-performing-countries/canada-overview/canada-teacher-and-principal-quality/.

Council of Ministers of Education, Canada. 1997. *Common Framework of Science Learning Outcomes K to 12*. Toronto, ON: Author, http://www.cmec.ca/9/Publications/index.html.

———. 2008. *The Development of Education Reports for Canada*. Toronto, ON: Author, http://www.cmec.ca/Publications/Lists/Publications/Attachments/122/ICE2008-reports-canada.en.pdf.

———. 2013. *Measuring up: Canadian Results of the 2012 OECD PISA Study*. Toronto, ON: Author, http://www.cmec.ca/docs/pisa2012/PISA2012_Highlights_EN_Web.pdf.

———. 2014. *Pan-Canadian Assessment Program (PCAP)*. Toronto, ON: Author, http://www.cmec.ca/302/Programs-and-Initiatives/Assessment/Pan-Canadian-Assessment-Program-%28PCAP%29/PCAP-2013/index.html.

Crocker, Robert, and David Dibbon. 2008. *Teacher Education in Canada: A Baseline Study*. Kelowna, BC: Society for the Advancement of Excellence in Education.

Falkenberg, Thomas. 2010. "Introduction: Central Issues of Field Experiences in Canadian Teacher Education Programs." In *Field Experiences in the Context of Reform of Canadian Teacher Education Programs*, edited by Thomas Falkenberg and Hans Smits, 1–50. Winnipeg, MB: Faculty of Education of the University of Manitoba.

Fazio, Xavier, and Douglas D. Karrow. 2013. "Science Takes a Back Seat: An Unintended Consequence of Prioritizing Literacy and Numeracy Achievement." *Education Canada*, 53 (3), http://www.cea-ace.ca/education-canada/article/science-takes-back-seat.

Gambhir, Mira, Kathryn Broad, Mark Evans, and Jane Gaskell. 2008. *Characterizing Initial Teacher Education in Canada: Themes and Issues*. Toronto, ON: Ontario Institute for Studies in Education, University of Ontario, http://www.oise.utoronto.ca/ite/UserFiles/File/CharacterizingITE.pdf.

Hurley, Stephen. 2015. "In the Beginning Was the Ballroom . . . The Changing Face of Professional Learning." *Education Canada*, 55 (1), http://www.cea-ace.ca/education-canada/article/beginning-was-ballroom.

MacDonald, Ronald. 2010. "Bridging the Theory-Practice Divide: Teaching Science Methods off Campus." In *Field Experiences in the Context of Reform of Canadian Teacher Education Programs*, edited by Thomas Falkenberg and Hans Smits, 261–73. Winnipeg, MB: Faculty of Education of the University of Manitoba.

Martin, Michael O., Ina V. S. Mullis, Pierre Foy, and Gabrielle M. Stanco. 2012. *TIMSS 2011 International Results in Science*. Boston College, MA: International Association for the Evaluation of Educational Achievement (IEA), http://timssandpirls.bc.edu/timss2011/downloads/T11_IR_Science_FullBook.pdf.

Milton, Penny. 2003. *Trends in the Integration of ICT and Learning in K–12 Systems*. Toronto, ON: Canadian Education Association, www.cea-ace.ca/sites/default/files/cea-2003-ict-integration.pdf.

Norris, Stephen P., and Linda M. Phillips. 2003. "How Literacy in Its Fundamental Sense Is Central to Scientific Literacy." *Science Education*, 87 (2): 224–40. doi:10.1002/sce.10066

Olson, Joanne K., Christine D. Tippett, Todd M. Milford, Chris Ohana, and Michael P. Clough. 2015. "Teacher Preparation in a North American Context." *Journal of Science Teacher Education*, 26 (1): 7–28. doi:10.1007/s10972-014-9417-9

Ontario Ministry of Education. 2007. *The Ontario Curriculum: Grades 1–8 Science and Technology*. Toronto, ON: Author, http://www.edu.gov.on.ca/eng/curriculum/elementary/scientec18currb.pdf.

Organisation for Economic Co-operation and Development (OECD). 2013a. *OECD Skills Outlook 2013: First Results from the Survey of Adult Skills*. Paris: OECD Publishing. doi: 10.1787/9789264204256-en

———. 2013b. *TALIS 2013 Results: An International Perspective of Teaching and Learning*. Paris: OECD Publishing. doi:10.1787/9789264196261-en

Osborne, Jonathon, and Sarah Hennessy. 2003. *Literature Review in Science Education and the Role of ICT: Promise, Problems and Future Directions* (Future Lab Series Report 6). Bristol: Future Lab, http://archive.futurelab.org.uk/resources/documents/lit_reviews/ Secondary_Science_Review.pdf.

Partnership for 21st Century Skills. 2009. *P21 Framework Definitions*. Washington, DC: Author, http://www.p21.org/storage/documents/P21_Framework_Definitions.pdf.

Reiti, John. 2014. "Ontario Announces $150 Million Investment for iPads in the Classroom." *CBC News*, September 4, 2014. http://www.cbc.ca/news/canada/hamilton/ontario-announces-150-million-investment-for-ipads-in-the-classroom-1.2755755.

Statistics Canada. 2014. *The Output of Educational Institutions and the Impact of Learning*. Ottawa, ON: Author, http://www.statcan.gc.ca/pub/81-604-x/2014001/hl-fs-eng.htm.

Tippett, Christine D. 2011. "Scientific Literacy: Disciplinary Literacy and Conceptual Understanding." Paper presented at the Eighth Biennial Conference of the European Science Education Research Association, Lyon, France, September 2011.

Trent University School of Education. 2014. *Teacher Education Stream*. Peterborough, ON: Author, http://www.trentu.ca/education/artsandscience.php.

University of Lethbridge Faculty of Education. 2014. *Combined Degree Programs*. Lethbridge, AB: Author, http://www.uleth.ca/education/programs/undergraduate-studies/combined-degrees.

University of Regina Faculty of Education. 2013. *Secondary Programs*. Regina, SK, http://www.uregina.ca/education/programs/secondary.html.

Further Reading

Falkenberg, Thomas, and Hans Smits, eds. 2010. *Field Experiences in the Context of Reform of Canadian Teacher Education Programs*. Winnipeg, MB: Faculty of Education of the University of Manitoba.

Falkenberg, Thomas, and Hans Smits, eds. 2011. *The Question of Evidence in Research in Teacher Education in the Context of Teacher Education Program Review in Canada*. Winnipeg, MB: Faculty of Education of the University of Manitoba.

Lewthwaite, Brian and Rick Wiebe. 2011. "Fostering Teacher Development to a Tetrahedral Orientation in the Teaching of Chemistry." *Research in Science Education*, 41 (5): 667–89. doi: 10.1007/s11165-010-9185-2

Milford, Todd M. 2014. "Canada: British Columbia." In *Issues in Upper Secondary Science Education: Comparative Perspectives*, edited by Barend Vlaardingerbroek and Neil Taylor, 47–66. New York: Palgrave Macmillan.

Steele, Astrid. 2011. "Beyond Contradiction: Exploring the Work of Secondary Science Teachers as They Embed Environmental Education in Curricula." *International Journal of Environmental & Science Education*, 6 (1): 1–22.

5

The Netherlands

Lesley de Putter-Smits and Jan van Driel

Teaching Science in the Netherlands

Backdrop

The Netherlands, a country with a population of 16.8 million, is home to people of many different origins, including significant minorities from Turkey, Morocco, Indonesia, Germany, Surinam, the Dutch Antilles, Poland, and Belgium. Children can begin attending government-provided education from four years of age but are not compelled to attend until they reach age five. From then on, they must remain in school until age 18, which marks the end of compulsory schooling. Most schools are run on government funding. Private schools, of which there are few in the Netherlands, cater for gifted children, children failing in standard education, and adult education. Home schooling is extremely rare. Kindergarten is integrated into primary schooling. Before kindergarten, children can attend preschool education, where knowledge of the Dutch language and social behavioral norms are the main goals. Preschools are found in larger cities where many people from different origins live and in rural areas where dialects such as Friesian are spoken.

Primary education encompasses Grade 1 (four-year-olds) through to Grade 8. During this time, children are taught the Dutch language (grammar, reading, writing), arithmetic, arts and crafts, physical education, and "world orientation," a subject that each school can individually interpret as comprising biology, geography and Earth science, history, social studies, early physics and chemistry, and so on (Greven and Letschert 2006). Nowadays, most schools include English language as a subject from Grades 7 and 8. Many schools also operate a social skills program. Primary schools can choose whether or not to have their Grade 8 students sit national examinations. Although these examinations are optional, about 85 percent of all primary schools choose to participate. The examinations provide the children and their parents with an indication of the kind of secondary education to follow. Teachers in the non-participating schools provide Grade 8 students and their parents with recommendations on which secondary courses students should follow.

Separate primary schools for special needs children are a standard part of the Dutch education system. Since 2014, greater numbers of special needs children are being mainstreamed into the general education system in an attempt to cut back on education costs.

Secondary education in the Netherlands falls into two broad categories: basic vocational education (27 percent of students) and general education (73 percent). General education is divided into three tracks: one that leads to vocational education after Grade 12 and serves almost 27 percent of students, one that leads to higher vocational education after Grade 13 and serves another 27 percent of students, and one that leads to university education after Grade 14 and serves about 19 percent of students.

All general secondary education concludes with a national examination in May of the terminating year (Grade 12, 13, or 14), although not all subjects taught are included—physical education and social studies are examples. The summative score for a specific subject is based on both internal school assessments and the external examination, with weightings of 50 percent each. Teachers develop and implement the school-based assessments; when and how these take place varies. To ensure the reliability of these assessments, the average school mark for a subject cannot differ more than 0.5 (on a 1 to 10 scale) from the average mark for the subject in the national examinations. Schools that do not or cannot comply with this stipulation are visited by the government inspectorate of education. These schools are expected to present a valid explanation for the difference in marks to avoid stringent inspection or cautionary measures. Students who pass the national examinations have right of entry to vocational education (Grade 12: VMBO-t diploma), higher vocational education (Grade 13: HAVO diploma), or university education (Grade 14: VWO diploma).

School Science

After completing the world orientation subject in primary school, all learners follow a basic curriculum in science subjects up to Grade 11. Biology and a subject called "science," which incorporates concepts from physics, chemistry, and technology-oriented topics, are commonly taught in Grade 9. These two subjects are succeeded by the single-science subjects of physics, chemistry, and biology in Grades 10 and 11. During lower secondary education, schools can choose the kind of science they wish to teach. Some schools prefer a more general "science and technology" subject, while others decide to teach biology, chemistry, and physics as separate subjects from Grade 9 onward. In 2003, a number of schools known as technasia introduced a new science and technology subject, *onderzoeken en ontwerpen* (i.e., research and design, abbreviated as O&O). Technasia use O&O as a means of delivering science and technology to Grades 9 to 11 students in a way that "speaks" to them by requiring them to engage in research and design projects. Each project is based on an authentic "real-world" issue and includes collaboration between students and people in a company or enterprise. Every technasium has a science and technology lab where students can work on their projects. After Grade 11, O&O is no longer compulsory, but students can elect to continue on with a science-oriented curriculum. The teachers who teach O&O come from a variety of disciplinary backgrounds, and they qualify to teach the subject by participating in a special course developed by the Technasium Foundation. Over the last 10 years, 85 secondary schools have received technasium certification.

Upper secondary education in the Netherlands is defined as the school years from Grade 11 through to graduation in Grade 13 or 14 (HAVO or VWO). The vocational Grade 12 VMBO-t is traditionally not considered part of upper secondary education. As indicated above, the school levels are distinguished not only by the number of years of secondary education a student completes but also by difficulty level and corresponding entry qualifications for further education.

At the end of their Grade 11 year, students enrolled in upper secondary education have to choose one of four national curricula. Two of these are science oriented ("nature and society" and "nature and technology"). The third encompasses economics and finance, and the fourth languages and culture. Only those upper secondary students taking a science-oriented curriculum study advanced physics, chemistry, and/or biology and/or Earth sciences and research and design. Mathematics is compulsory for all students up to the final examination, which they have to pass.

Almost 50 percent of secondary schools in the Netherlands offer the new subject "nature, life, and technology" (NLT) to upper-level science students. This non-compulsory subject was added to the curriculum in 2007 as an initiative to promote interest in science (Stuurgroep NLT 2007). NLT combines at least two science disciplines per project and requires students to engage with science content at a much deeper level than the separate science subjects do. The syllabus consists of modules developed and tested by teams of science teachers and then approved by a national committee. Schools are free to choose from the range of modules available. NLT is preferably taught by a team of teachers from different science subject backgrounds to ensure in-depth coverage of the content in the various modules. NLT is taught mainly in Grades 12 and 13 and does not have a national examination. Computer science (programming) is taught at the upper levels of 60 percent of secondary schools. This subject is similar to NLT, can only be chosen as part of a science-oriented curriculum, and also has no national examination.

In 2013, the Netherlands implemented new curricula for biology, chemistry, and physics in upper secondary education. The previous curricula were considered to be overloaded, focused on isolated facts rather than presenting a coherent entity, and strongly content oriented. They were also seen as outdated and as not meeting the demands of modern society with respect to science education. The new curriculum thus had to contribute to scientific literacy, highlight the relevance of science for society, and improve learner motivation for science as a school subject as well as for further educational choices (Bertona 2012).

The common goal for the new curricula is to attract more students to science by ensuring that the content works within authentic or real-world contexts and so is more meaningful to students. Incorporating modern science into the different science curricula is seen as providing a more realistic image of the sciences in terms of both further education and the development of the modern world. An emphasis on scientific reasoning and on executing research and design projects is embedded in the new curricula. Introduction of the curricula has been carefully planned since 2003 and has striven to incorporate teachers' input on what to teach into both the formal curricula and the associated new curricular materials.

Teams of science teachers under the supervision of the Centraal Instituut voor Toetsontwikkeling (Central Institute for Test Development or Cito) are responsible for devising the national examinations for the separate subjects of biology, chemistry, and physics, while the College voor Toetsen en Examens (Institute for Testing and Examinations or CvTE) is responsible for administering them. The recent changes in the content of the science curricula have been incorporated into the national examinations, although assessing some competencies (e.g., research capabilities) through a written test remains difficult. These capabilities are mostly assessed internally through school-based assessments that include practical tests and research projects as well as cognitive tests on science content.

Teacher Academic and Professional Education and Training

According to Dutch law, teachers can start teaching without a teaching qualification but with the obligation to obtain one within two years after their first appointment. The Netherlands offers four teaching degrees: one for teaching in primary schools, one for teaching in secondary schools up to Grade 11, and two (a higher vocational education route and a university teaching degree) for teaching upper secondary school (Grades 12 to 14). Candidates for secondary education must undertake a specialized teaching degree for each subject they intend to teach. If, for example, someone wants to teach both chemistry and physics, he or she must obtain two separate qualifications.

Most science teachers in secondary education go through a higher vocational education track in biology, chemistry, or physics in order to obtain a teaching qualification for up to Grade 11. This preservice teacher training takes four years and includes relevant science content by subject, general pedagogy and pedagogy of science, internships in schools, and an educational research or design project.

Teachers who have obtained this qualification can follow a higher vocational track leading to the qualification that allows them to teach upper-level science (Grades 12 to 14). This track includes more in-depth science content for the subject chosen (chemistry, physics, or biology), specific teacher training for the upper level, and an educational research project. Teachers typically follow this track as an inservice program; it takes three years to complete.

Alternatively, students who have studied biology, chemistry, physics, or a closely related science subject successfully at the university level for at least three years (Bachelor's degree) or more (Master's degree) can enroll in a one-year science-teaching Master's program at university to obtain a teaching degree for upper secondary education in the corresponding science subject. This program involves internships, general pedagogy and pedagogy of science, and an educational research or design project. Content knowledge is not included given that students have already obtained a degree in a discipline strongly related to the school subject. If their degree is in a more remote discipline, such as civil engineering, the student is required to follow courses in the needed content subjects either before or during the teacher education program.

The reason why the Netherlands offers two Master's degrees for teaching upper secondary science is historical in origin. Higher vocational education in general is characterized by a slightly lower academic level in content taught because it is aimed at mastering a profession. University education is of a higher academic level because its focus is on mastery of academic skills. Today, the institutes for higher vocational education often call themselves "universities of applied science" in foreign communications.

Continuing Professional Development

In 2006, the Netherlands passed a law to ensure "continuing teaching quality" through continuous professional development of teachers and support staff (Ministry for Education, Culture, and Science 2006). School principals are expected to monitor their staff members' professional development needs, based on the teaching competencies defined by the Stichting Beroepskwaliteit Leraren (The National Teachers Assembly or SBL). The SBL has since been succeeded by the Onderwijscoöperatie, a federation of teacher organizations and teacher unions and the preferred interlocutor for the ministry. Each year, all teachers are expected to identify their personal learning goals and work accordingly on their professional development. Teachers and schools are free to choose from professional development programs offered by various educational institutes around the country, with choice based on the particular interests of the teacher or school. The government provides a fixed professional development budget per teacher per year through the school funding system.

In 2012, the Dutch government funded the establishment of regional science support centers where teaching institutes and science faculties collaborate to provide teacher support. The kind of support provided varies across the centers from closely cooperating networks of teaching institutes and high schools to a combined platform of teaching institutes that interested high schools or science teachers can turn to for professional development needs.

Many different suppliers provide professional development courses for secondary science education. Suppliers range from commercial training bureaus to government-funded professional development agencies and higher vocational education and university teacher training institutes. The government gives each school a lump sum for teacher professional development and lets the school board and individual teachers choose the kind of professional development they desire.

Options for professional development include studying up-to-date science content or science learning strategies, obtaining a higher teaching degree, and doing a PhD or even post-doctoral research. Teachers wanting to pursue the latter options can apply for a grant from a special government fund.

Science Teacher Professional Associations

The largest science teacher professional association in the Netherlands is the Nederlandse Vereniging voor Onderwijs in de Natuurwetenschappen (the Dutch Society for Education in the Sciences or NVON). The society presents the ministry with science teachers' views on curriculum changes, employment conditions, professional development opportunities, and so on. The society furthermore organizes yearly conferences, liaises with the CvTE about national examinations, arranges professional development programs in collaboration with educational institutes, and supports new ideas and developments in science education.

Other associations focus on one specific science subject. For example, biology teachers have at hand the Netherlands Institute of Biology (NIBI), chemistry teachers the Royal Dutch Chemistry Association (KNCV), and physics teachers the Dutch Physicists Association (NNV). The Freudenthal Institute for Science and Mathematics Education (FISME) caters for chemistry and physics education. These agencies organize well-attended yearly conferences, publish magazines and websites showcasing new findings in the subject field, and provide teaching ideas and tips for the specific subject.

Issues in Upper Secondary Science Teacher Quality

Teacher Supply

Science teachers are recruited from two main backgrounds. The first set of teachers come from the regular teacher training programs (higher vocational and university) and so are mainly young teachers at the beginning of their working career. Others are drawn from other occupations, such as engineering. These teachers choose education as a career for various reasons, such as a desire to change direction in life or to experience new opportunities. Teachers entering the profession are of varying ages.

The need for teachers in upper-level mathematics (and Dutch language) is still growing, given that many teachers are approaching retirement age

(Lubberman et al. 2013). The demand for physics and chemistry teachers seems to be stable. For other science subjects, in particular biology, supply surpasses demand. However, there is a shortage of teachers with Master's degrees, compulsory for anyone teaching Grades 12 to 14, particularly in physics and mathematics.

Surveys of science students reveal that many of these students harbor negative images of teaching as a profession: university students consider teachers as being relatively low paid, and they tend to think that becoming a teacher seriously limits other career options ("once a teacher, always a teacher"). In order to increase the number of teachers, the Dutch government has implemented various initiatives, most of which endeavor to make teacher education more flexible in terms of candidate selection and program structure. As an example, the Eerst-de-klas (Teach First) initiative, copied from British and US initiatives, draws 30 to 40 young science graduates to the teaching profession each year. These intensive programs combine a teaching job, industrial working experience, and teacher training in order to give young professionals a head start in both worlds. To encourage people to consider teaching as a profession, the government introduced a new salary scale in 2005 and ran a series of television commercials promoting teaching as a profession. These initiatives seem to have increased the influx of student teachers.

Professional Education and Training Issues

In 2013, the government published plans for the teaching profession in *Lerarenagenda 2013–2020* (Teacher Plan). The plan includes measures to counteract teacher drop-out, create options for combining teaching and working in industry, and ensure opportunities for continued professional development. *Lerarenagenda 2013–2020* also includes plans for enticing higher-caliber students to enroll in teacher education programs and to increase the quality of teacher education (Ministry for Education, Culture, and Science 2013).

The subject knowledge of students enrolling in teacher training programs is one of the first issues being addressed. Higher vocational teacher training institutes adhere to a common content knowledge base for teachers (HBO-Raad 2009). In contrast, because of differences in the kind, depth, and width of content knowledge studied in each university, a prospective teacher wishing to attain a university Master's degree in teaching a science subject has to meet varying requirements in order to enter a teacher training program. The people running university teaching programs for physics and mathematics are currently discussing possibilities for broadening the scope of university studies that give access to a university Master's degree in teaching. The goal is to create not only an acceptance policy but also generally available courses that cover the subject knowledge potentially missing from a prospective student's curriculum vitae.

Many unqualified science teachers follow a part-time inservice teacher training program alongside their jobs. Because a permanent appointment depends on having a teaching degree but life demands an income, inservice student teachers experience the tensions of juggling their job, their family's needs, and the demands of their educators. These student teachers often take twice as long as usual to complete a degree.

Much of the current research on teacher education in the Netherlands focuses on general pedagogical issues, such as student behavior and classroom management. Recently, because of the increase in computer-based teaching in the Netherlands and the ubiquitous availability of the Internet, the Dutch *VELON* journal for teacher educators devoted a special issue to ICT use during teacher training. In general, the conclusions of the research presented in the journal were disheartening. In the main, ICT use in teacher training is still poor. Initiatives suggested included working with teams of student teachers to create curriculum materials with an emphasis on using ICT, stimulating student teacher creativity in using ICT, and creating more opportunities and showcasing more possibilities for these uses (Martens, Van den Berg, and Vanderlinde 2014).

Some research is devoted to science teacher training, the findings of which are incorporated into the university teacher training programs or used to inform effective secondary science teaching. Research into the pedagogical content knowledge (PCK) of preservice chemistry teachers, for instance, reveals the necessity of combining lecture-based training on a specific (mis)conception with teaching the concept at hand in class. De Jong, van Driel, and Verloop (2005) found that preservice teachers' PCK increased significantly when they experienced this method of training.

Teachers in the Netherlands usually have a permanent employment contract with a school, which means they remain employed at that school until retirement age (which is presently slowly rising from 65 to 67 years of age in the Netherlands). Teachers sometimes change schools for economic or motivational reasons, but they are rarely fired. This high degree of job security and autonomy at the school level tends to create distinct institutional cultures. While the education law of 2006 gives school boards some power to mandate teachers to join inservice training programs, coercion is rarely needed to motivate teachers to participate in professional development courses.

Issues at the Chalkface

In the Netherlands, science teaching is increasingly transforming from traditional teaching to more constructivist ways of teaching. Problem-based education, context-based education, and various forms of student-directed active learning are promoted in teacher education, the teaching materials accompanying the new science curricula, magazines and websites devoted to science teaching, and so on.

In the late 1990s, when interest in science studies was waning and a great many scientists in the industry were nearing retirement age, cooperation between high school education and industry was formalized through the establishment of JETNET, an organization initially consisting of five major Dutch technological companies such as Shell and Philips. JETNET's goal was to stimulate more interest in science among young people, an initiative that, according to recent research, has been successful (Vennix, den Brok, and Taconis 2014). As a spin-off, more practical, authentic scientific practices are now available to teachers as part of their classroom practice. Teachers now find it easier to visit companies and to invite people from industry to give students an authentic picture of science and technology and thereby heighten their enthusiasm for science. This impulse to increase the population of scientists has given teachers greater awareness of the reasons why students should learn science and, moreover, *how* they should learn it.

Since 2005, schools in the Netherlands have been able to apply for a grant to promote science education in their schools. They do this through the Platform Bètatechniek (Science and Technology Platform), funded by the Ministry of Education, Culture, and Science. Schools use this grant to improve their laboratory facilities and to develop new science teaching programs with the aim of attracting more students from Grade 11 onward to study science.

The new curricula for the sciences in upper secondary education have adopted a context–concept approach, which means they have shifted from a content-oriented perspective to a perspective favoring continuous interaction between contexts and concepts and where ability to use knowledge in different contexts is of primary importance. Attainment targets now focus on the application of knowledge. The contexts are the starting point for the exploration and application of science and function as a bridge between students' perceptions of the world and abstract scientific concepts. Contexts pertain to the theoretical, experimental, social, or vocational fields and are chosen to meet curricular requirements such as the development of knowledge and skills, the experimental nature of science, scientific literacy, the position of science in Dutch society, and the relevance of science for different professions. The associated teaching materials currently available vary widely in how they interpret using a context as the starting point for exploration of concepts. In some materials, for example, contexts are only used to illustrate concepts.

Within the new science curricula, teachers have the option of using free online materials, created and trialed by their peers, or published textbooks. Some schools have decided to use online textbooks only ("iPad schools"). Advances in technology have made it easier for schools to use teaching materials previously difficult to obtain. For example, YouTube hosts many science movies and lectures, while the web in general provides teachers with opportunities to use web-based lectures and online learning environments. The web also offers applets that students can use to study and clarify different scientific concepts. In short, these developments have enabled teachers to substantially expand their pedagogical resource base.

Many science teachers in the Netherlands have taken up these opportunities and now use the web as a major teaching resource. Some go as far as to completely change their classroom procedures by having students do only experiments, exercises, and projects in class and then access explanations of concepts at home by watching short instructional videos such as those on the YouTube channels of Dutch science teachers (see the further reading section at the end of this chapter for examples). Another trend is for teachers to gather, arrange, and develop their own teaching materials based on their knowledge and experience, now made easier to use in classrooms through modern ICT means, such as interactive whiteboards and iPads.

Interest in students learning science through research-type laboratory work (as opposed to the traditional textbook approach) is increasing. Some schools have built full science laboratory facilities for students to work in and have employed technicians as support staff. A well-equipped science and technology workplace is obligatory in technasium schools. Not all schools have the means or interest to attain these types of facilities, but most science teachers consider them desirable because they agree that students should experience science as it is conducted in the "real world."

Some schools, however, have reacted to this constructivist way of teaching by electing to remain solidly cognitivist, with rule-based ex-cathedra teaching and so-called "cookbook" experiments. Both types of schools are popular, and families in the densely populated areas of the Netherlands usually have no problem finding a school type of their choosing. Because university entry requirements are based on the subjects taken in the final examinations and not on how they were taught, the type of school a student attends does not, in principle, affect his or her options for further education.

The number of hours teachers spend actually teaching has diminished somewhat since 2003 from 28 to 26 teaching hours for a full-time teaching position (in the Netherlands, 40 hours of work a week constitutes full-time employment). Teaching is still considered a stressful occupation, even though school holidays continue to be holiday periods for secondary education teaching staff. However, as of 2015, primary teachers have had their holidays reduced by one week so that they can spread out their workload across a slightly longer period of time each school year.

Schools are bulk-funded, which means each school board can decide how to use its funding allocation in its school. Some boards apportion money in a way that means the teaching load in their respective schools is very high. In this instance, boards probably decided to have large classes (over 30 students per class) so that other school activities could be pursued. When teachers, including science teachers, are expected to teach large classes, it is not the quality of the teacher that should be questioned but the time available to meet each student's learning needs.

Teaching in every school type and subject has become more challenging since the law on "inclusive education" came into effect at the beginning of January 2014. Previously, children with special needs attended special schools or were home schooled. Most teachers have not had the training required to cater for the needs of these children, a situation that is likely adding to the stress levels of teachers throughout the school sector. Meeting the learning needs of students in large classes (30-plus students in some cases), let alone classes where students vary widely in ability and learning acumen, is not an easy task for any teacher and is particularly difficult for teachers of upper-level science because of the type of pedagogy that the new curricula require. Creating an effective learning environment in classes such as these is expected to become an important issue in science education over the coming years.

Continuing Professional Development Issues

As an initiative complementary to the envisaged new science curricula, some 20 science teachers were given the opportunity to complete doctoral study by researching various topics associated with the new subjects. The topics chosen included considering teacher professional development with respect to innovative teaching of the new curricula, designing study modules, looking at new ways of educating students through the new curriculum, and examining coherence across the different curricula (Knippels, Goedhart, and Plomp 2008).

The findings of these studies have provided insight into ways of helping teachers adapt to the new curricula. For example, having studied the pedagogical approaches of science teachers in biology, chemistry, NLT, and physics, De Putter-Smits et al. (2012) found that in the subject areas of chemistry and biology Dutch classrooms had moved toward a more context-based approach before the new curricula had even been implemented. Both Visser et al. (2012) and Dam (2014) devised professional development courses and tools that science teachers could use to help them teach the new curricula. A related project showed that teachers who were part of the teams designing new materials for the revised curricula reformed their teaching toward context-based education, whereas teachers who only used the new materials did not (De Putter-Smits 2012).

Recently, more grants have become available for research on professional development for science teachers vis à vis the new science curricula, with that development focused not only on the curricular content but also the required pedagogy (i.e., context-based education). In addition, science support centers are providing universities and higher vocational education institutes with grants to help them develop and provide professional development courses for teachers on the new curricula. The contents of those courses are being informed by the findings of the above-mentioned research.

Although research shows that one-off workshops are not effective in changing teaching practice, school boards still tend to buy generic professional development courses for their teaching workforce. The learning communities envisaged by, among others, Shulman and Shulman (2004) are gaining ground, however, in the professional development portfolios of higher vocational and university teacher training institutes. For example, both types of institute have set up educational design teams and professional learning communities. These teams support science teachers as they endeavor to incorporate, test, and evaluate the particulars of the new curricula in their school practice. Research conducted by Coenders (2010) shows that teachers who elaborated on the new chemistry curriculum by structuring and designing teaching materials in a learning community were able to accommodate the changed curriculum more quickly than other teachers. The teams are usually headed by an institute for teacher education and attract science teachers from different high schools. The process also has the obvious advantage of facilitating collaboration and exchange among science teachers from different schools.

Trends and Developments in Upper Secondary Science Teacher Quality

The Netherlands' teacher professional development law (Wet BIO) is based on teaching competencies that the SBL established in 2006 (see earlier in this chapter). Although efforts are being made to update these competencies, most training programs and professional development initiatives are based on them. The competencies include interpersonal attributes, general pedagogical knowledge, content knowledge, pedagogical content knowledge, classroom management skills, and professional behavior.

In 2012, the Netherlands began, as part of its efforts to promote teacher professional development, a teacher register (*Registerleraar*) to which teachers can upload their CVs and list their professional development status. Agencies offering professional development courses can also register with *Registerleraar* as a guarantee of course quality. However, because teachers are not obligated to register, many have not done so. Registration will be compulsory from 2017, though, as stipulated by *Lerarenagenda 2013–2020* (Teacher Plan). As yet, registering professional development activities on the teacher register is not working as well as it should, thus diminishing its usefulness for both teacher trainers and teachers.

Currently, 72 percent of teachers drop out of the teaching profession within their first five years of training/teaching (Dekker 2013). About 40 percent of these individuals leave during their teacher education, usually because of a mismatch between expectations and the realities of the course of study or the profession, or simply a lack of motivation to continue. About 23 percent of each cohort that graduates from teacher education take up a profession outside education, while 9 percent quit the profession for reasons such as workload and lack of support.

A government-funded national induction program that aims to keep beginning teachers in the profession and support them to improve their performance is now in place. Teachers with up to three years of working experience are invited to participate in the program, which is led by educational institutes, both higher vocational and university, in close collaboration with high schools. Teacher coaches from high schools are trained to coach beginning colleagues in their schools. Coaching of new teachers is conducted through video coaching, peer feedback systems, and pedagogical workshops. The model is based in part on research by Beijaard, Buitink, and Kessels (2010) and Pillen, Beijaard, and den Brok (2013), the findings of which highlight important success elements in teacher induction programs, such as class management support (pedagogical knowledge in practice), emotional support and feedback, and fostering the growth of a professional identity. Taken together, these measures aim to increase the attraction of teaching science as a career and to improve both the quality and the quantity of qualified science teachers in the Netherlands.

References

Beijaard, Douwe, Jaap Buitink, and Chantel Kessels. 2010. "Teacher Induction." In *International Encyclopedia of Education*, edited by Penelope L. Peterson, Barry McGraw, and Eva Baker, 3rd ed., vol. 7, 563–68. Oxford: Elsevier Scientific Publishers.

Bertona, Cris. 2012. "Defining the Structure and the Content of a New Chemistry Curriculum in the Netherlands." In *Making It Tangible: Learning Outcomes in Science Education*, edited by Sascha Bernholt, Knut Neumann, and Peter, Nentwig, 457–73. Münster: Waxmann.

Coenders, Fer. 2010. "Teachers' Professional Growth During the Development and Class Enactment of Context-Based Chemistry Student Learning Material." PhD diss., Twente University, http://purl.utwente.nl/publications/71370.

Dam, Michiel. 2014. "Making Educational Reforms Practical for Teachers: Using a Modular, Success-Oriented Approach to Make a Context-Based Educational Reform Practical for Implementation in Dutch Biology Education." PhD diss., Leiden University Graduate School of Teaching, Leiden, The Netherlands, http://hdl.handle.net/1887/25806.

De Jong, Onno, Jan H. van Driel, and Nico Verloop. 2005. "Preservice Teachers' Pedagogical Content Knowledge of Using Particle Models in Teaching Chemistry." *Journal of Research in Science Teaching* 42 (8): 947–64.

Dekker, Sander. 2013. "Letter to President of House of Representatives, 13 March 2013 on Boosting Teacher Numbers in Subjects with Teacher Shortages" (in Dutch). The Hague: Ministry for Culture, Education, and Science, http://www.rijksoverheid .nl/ bestanden/documenten-en-publicaties/kamerstukken/2013/03/14/kamerbrief -over-invulling-impuls-leraren-tekortvakken/kamerbrief-over-invulling-impuls -leraren-tekortvakken.pdf.

De Putter-Smits, Lesley G. A. 2012. *Science Teachers Designing Context-Based Curriculum Materials: Developing Context-Based Teaching Competence.* PhD diss., Eindhoven Technical University, Eindhoven, The Netherlands.

De Putter-Smits, Lesley G. A., Ruurd Taconis, Wim Jochems, and Jan van Driel. 2012. "An Analysis of Teaching Competence in Science Teachers Involved in the Design of Context-Based Curriculum Materials." *International Journal of Science Education* 34 (5): 701–21.

Greven, Jan, and Jos Letschert. 2006. *Kerndoelen Primair Onderwijs* [Core objectives of primary education]. The Hague: Ministry for Education, Culture, and Science, http://www.slo.nl/primair/kerndoelen/ Kerndoelenboekje.pdf/.

HBO-Raad vereniging voor hoge scholen [Association of Colleges]. 2009. *Kennisbasis, Biologie, Natuurkunde, Scheikunde, Techniek, Wiskunde* [Knowledge base for biology, physics, chemistry, engineering, mathematics]. The Hague: Author, http://www.fi. uu.nl/publicaties/literatuur/2009_ kennisbasis_ lerarenopleiding_vo _betastudies.pdf.

Knippels, Marie-Christine, Martin J. Goedhart, and Tjeerd Plomp. 2008. "Docenten in Onderzoek—Het DUDOC-programma [Teachers doing research—the DUDOC program]." *Tijdschrift voor Didactiek der β-wetenschappen* 25: 51-70.

Lubberman, Jos, Nico Van Kessel, Menno Wester, and Ardi Mommers. 2013. *Rapport Arbeidsmarktanalyse Voortgezet Onderwijs* [Labour market analysis report: Secondary education]. Heerlen: Voion.

Martens, Rob, Jeroen Thys, Ellen Van den Berg, and Ruben Vanderlinde. 2014. "Redactioneel" [Editorial]. *Tijdschrift voor lerarenopleiders: themanummer ICT in de lerarenopleiding* 35 (4).

Ministry for Education, Culture, and Science. 2006. *Wet op de beroepen in het onderwijs* [Law on teacher professional development]. The Hague: Author, http://www.wetboek-online.nl/wet/Wet%20op%20de%20beroepen%20in%20het%20onderwijs.html.

——. 2013. *Lerarenagenda 2013-2020* [Teacher plan 2013-2020]. The Hague: Author, www.lerarenagenda.nl.

Pillen, Marieke, Douwe Beijaard, and Perry den Brok. 2013. "Tensions in Beginning Teachers' Professional Identity Development, Accompanying Feelings and Coping Strategies." *European Journal of Teacher Education* 36 (3): 240–60. doi: 10.1080/02619768.2012.696192

Shulman, Lee S. and Judith H. Shulman. 2004. "How and What Teachers Learn: A Shifting Perspective." *Journal of Curriculum Studies* 36 (2): 257–71.

Stuurgroep NLT. 2007. *Contouren van een nieuw Bètavak* [Contours of a new science course]. Enschede: Author, http://www.betanova.nl/downloads/Visiedocumenten/Contouren_20van_20een_20nieuw_20betavak.pdf/.

Vennix, Johanna, Perry den Brok, and Ruurd Taconis, R. 2014. "Outreach and Student Motivation in the K-12 STEM Field: Teachers' Roles and the Connection Between Industry and School." Paper presented at the European Educational Research Association Conference 2014: The Past, the Present, and the Future of Educational Research, http://www.eera-ecer.de/ecer-programmes/conference/19/contribution/31326/.

Visser, Talitha C., Fer G. M. Coenders, Cees Terlouw, and Jules M. Pieters. 2010. "Essential Characteristics for a Professional Development Program for Promoting the Implementation of a Multidisciplinary Science Module." *Journal of Science Teacher Education* 21 (6): 623–42.

Further Reading

Bybee, Rodger W. and Bruce Fuchs. 2006. "Preparing the 21st Century Workforce: A New Reform in Science and Technology Education." *Journal of Research in Science Teaching* 43 (4): 349–52. doi: 10.1002/tea.20147

Driessen, H. P. W. and H. A. Meinema. 2003. *Chemistry Between Context and Concept: Designing for Renewal.* Enschede: Netherlands National Institute for Curriculum Development, http://www.slo.nl/voortgezet/tweedefase/vakken/scheikunde/english_version_chemistry_between/986_Chemistry_20between_20context_20and_20concept. pdf/.

Examples of Youtube channels by Dutch science teachers (and students!)

- SiegerKooij: https://www.youtube.com/user/scheikundelessen
- PieterPalsma: https://www.youtube.com/user/ACNatuurkunde
- Students and Teacher Scala Rietlanden: https://www.youtube.com/user/biologielessen
- Hiele brothers: https://www.youtube.com/user/WiskundeAcademie

Reports on innovations in the different science curricula are available through http://www.betanova.nl/documentatie/achtergrond/, while the general content that teachers need to master in order to teach secondary school science subjects (Grades 9 to 11) can be found in the following publication:

HBO-raad vereniging voor hoge scholen [Association of Colleges]. 2011. *Generieke Kennisbasis Tweedegraads lerarenopleidingen* [Generic knowledge for secondary teacher training]. The Hague: Author, https://10voordeleraar.nl/documents/kennisbases_bachelor/kb-generiek.pdf.

6

Japan

Toshihide Hirano

Teaching Science in Japan

Backdrop

Japan is an island nation of Eastern Asia. About 70 percent of its land is forested and mountainous, and its 127 million citizens (98.5 percent of whom are ethnic Japanese) live mainly in the habitable coastal areas of the country's five main islands. Japan is administratively divided into 47 prefectures, and each prefecture is divided into cities, towns, and villages. The Ministry of Education, Culture, Sports, Science, and Technology (MEXT) administers the nation's education system, with this work including educational planning and development of prescriptive "courses of study." Prefectural boards of education administer the plans and budgets for institutions such as upper secondary schools and educational training centers. They also oversee the personnel affairs of public school teachers (recruitment, replacement, and promotion). Municipal boards of education administer the compulsory school sector (primary schools and lower secondary schools).

Japan's modern national school system was established in 1872 (Ministry of Education, Science, and Culture 1980). Over subsequent years, Japan gradually extended the structure of this system with reference to the educational systems and trends of Western countries but always modifying those models to suit Japanese custom. After World War II, the Basic Act on Education and the School Education Law were promulgated and put into effect in 1947. This legislation brought in schooling based on the United States' 6-3-3-4 system of education. Nursery schools and one- to three-year kindergartens for preschool children were also established at this time. The proportion of students enrolled in private-sector institutions within the compulsory school sector is less than 10 percent of the total school population, but more than 60 percent in kindergartens and about 30 percent in upper secondary schools.

Upper secondary schools were first established in 1948. They offered three-year full-time and part-time courses. In 1961, correspondence courses became part of the system. The proportion of the eligible age group attending upper secondary school steadily increased year by year from 43 percent in 1950 to reach 52 percent in 1955, 71 percent in 1965, and 92 percent in 1975. Currently, over 97 percent of lower secondary school students advance to general or vocational upper secondary schooling at a ratio of about three to one. Public upper secondary schools select students on two bases: student results on achievement tests set by prefectural boards of education of five main subject areas including science, and the records of each applicant presented in the credentials or other documents submitted by lower secondary schools. The Act on Free Tuition Fees at Public High Schools and the High School Enrollment Support Fund, which was passed and enacted in 2010, ensures free tuition at public upper secondary schools and established the tuition payment support fund system for private school students.

Japan's courses of study set broad standards for all schools, from kindergartens through to upper secondary, in order to ensure a uniform delivery and quality of education throughout the country. Schools organize their programs for all subjects and special activities according to the standards, which are generally revised every ten years. The most recent revision for compulsory schools took place in 2008, and for upper secondary schools in 2009 (Ministry of Education, Culture, Sports, Science, and Technology 2009a). At that time, MEXT canceled the aim of "cram-free education" (a policy that saw science content and science study time drop over the next two decades) and shifted the focus to developing students' logical reasoning ability and motivation for learning. This change in focus was a response to weaknesses arising from Japanese students' performance in international surveys of educational achievement such as the Trends in Mathematics and Science Study (TIMSS) and the Programme for International Student Assessment (PISA).

Today's courses of study are intent on nurturing a "zest for life" in students in accordance with the educational principles expressed in the revisions to the Basic Act on Education in 2006. The courses of study also reflect Japan's commitment to enriching curricular content and boosting by 10 percent the number of classes in major subject areas, including science. Strong heed is being given to international acceptance and academic consistency of curricular content and to the balance between acquiring basic knowledge and fundamental skills and fostering the ability to think, make decisions, and express oneself through engagement with the spiral curriculum, observation/experimentation, and assignments.

The School Education Law created the textbook approval system in 1947, in which non-governmental publishers of textbooks must submit them for official approval by MEXT. While these books must meet the requirements of the courses of study, publishers have the freedom to include their own learning methods and ideas in the material. The municipal boards of education (for public compulsory schools) or the schools themselves decide which books to adopt from the lists of approved texts. This system also serves as a means of standardizing the quality of school education in Japan.

School Science

At compulsory schools, school science is taught as general science that unifies physical science, biology, and Earth science. The subject area named *rika* in Japanese was introduced into higher elementary schools in 1886 and set its objectives in terms of combining "Western-based science learning," a scientific mind-set, scientific knowledge, and the "Japanese spirit" of loving harmony with nature. From 1947 to 1991, children began learning *rika* on entering Grade 1 of elementary school. However, in 1992, children only began learning this subject from Grade 3 on because it was at this time that the new integrated subject area "life environment studies" was introduced for children in Grades 1 and 2. *Rika* is taught mostly by homeroom teachers. Otherwise, it is taught by subject-specialist teachers in elementary schools and by subject teachers in lower secondary schools.

MEXT has been carrying out national assessments of sixth- and ninth-graders' academic ability in mathematics and Japanese since 2007. Science assessments were added in 2012 and 2015. The results of these assessments are used to inform educational policies and improve classroom teaching. The current courses of study provide teachers with clear information on the key scientific competencies that students are to acquire. This information substantially helps teachers both understand and manage the teaching of these competencies.

At the upper secondary level, the courses of study separate school science into ten subjects (Ministry of Education, Culture, Sports, Science, and Technology 2009b). A variety of basic- and advanced-level subjects are available under the umbrella of four discrete sciences: biology, chemistry, physics, and Earth science. Schools also have at hand the basic integrated science subject named "science and our daily life" (SODL) and the investigation-oriented, research-based "science project study" (SPS), both of which have been developed for students majoring in science. All Japanese upper secondary students must choose and then study three basic-level subjects from the four discrete sciences or one basic-level subject and SODL as a minimum requirement for graduation.

Students studying a general academic upper secondary course usually enter a humanities track or a sciences track before they reach Grade 12. Students in the sciences track usually take three or more advanced-level subjects from among the four discrete sciences because they are part of university entrance examination requirements. According to the Upper Secondary School Curriculum Survey conducted by MEXT in 2013 (2014b), around 80 to 90 percent of the upper secondary cohort at the time were enrolled in general academic courses of basic-level and advanced-level biology, chemistry, and physics, and about 50 percent were enrolled in basic Earth science. However, only 20 percent of upper secondary students were enrolled in advanced Earth science and SODL, and only 3 percent in SPS. MEXT has yet to release subject enrollment rates in advanced-level subjects, but when it does, the results will be released in the order of biology, chemistry, physics, and Earth science, in descending order. Every science subject is taught by subject teachers of science, usually selected according to their university specialization. However, Earth science and SODL are usually taught by science teachers who have some other area of specialization.

Teacher Academic and Professional Education and Training

Teacher training courses are conducted by MEXT-approved general universities and specialist teacher training universities. Students qualify for certification from either institution if they take the minimum necessary credits from the course subjects prescribed by the Education Personnel Certification Act and the Ordinance for Enforcement of the Education Personnel Certification Act. However, the course curricula that the two types of university provide are quite different in both the quantitative and qualitative senses.

The teaching certificates held by lower and upper secondary teachers of science are not distinguished with reference to discrete sciences. Instead, certificates are divided into classes that are given equivalence with degree levels. Thus, science teachers holding a first-class certificate will have earned a Bachelor's degree and acquired more than 59 credits from teaching and professional subjects, with the latter including at least two or four weeks of teaching practice experience at lower or upper secondary schools at the university-attached schools or cooperating public schools. Teachers holding an advanced certificate will have earned a Master's degree and added 24 credits for teaching or professional subjects to their first-class certificate. Coursework relating to the subjects these prospective teachers intend to teach builds up content knowledge and involves both lectures and experimental work for each of the four discrete sciences (physics, chemistry, biology, and Earth science). Professional subjects build up pedagogical knowledge and pedagogical content knowledge (PCK) and include practical teaching experience.

The prefectural boards of education award teacher certificates in accordance with the teacher certification legislation. Almost all lower secondary school science teachers have first-class regular certificates, but only 5 percent of them also have advanced certificates. However, 30 percent of upper secondary school science teachers have both first-class and advanced certificates.

Continuing Professional Development

Inservice professional development is provided mainly at prefectural training centers. One-year training programs for all newly hired teachers and all experienced teachers (i.e., teachers with ten or more years of teaching experience) are prescribed by the Special Act for Educational Personnel, as are short training courses for applicant teachers relating to subject teaching, student guidance, and school administration. Prefectural and municipal boards of education also provide limited opportunities for off-campus training in the form of educational research or graduate study. The National Center for Teacher Development also provides special training programs for lead teachers and subject inspectors.

In 2007, the government amended the Education Personnel Certification Act, and in 2009 MEXT implemented Japan's current system of renewing teacher certificates. The Act requires all teachers to renew their licenses every 10 years by attending at least 30 hours of the certificate-renewal courses provided by universities and other educational organizations approved by MEXT. The aim of the process is to keep teachers up to date with the latest in pedagogical knowledge and skills.

Science teachers are given opportunities to attend workshops and educational conferences provided by the national, prefectural, and municipal levels of science teachers' professional associations, educational institutions such as universities and science museums, and some companies. In addition to catering for preservice training, the schools attached to the national teacher training universities play an important role in inservice training. They conduct demonstration lessons followed by open discussion with attending teachers. This style of teaching classroom pedagogy is called *jugyo-kenkyu* (lesson study). The attached schools also host professional meetings of educational researchers.

Science Teacher Professional Associations

All public lower secondary school teachers are automatically members of municipal professional associations defined by subject area and school type. These include science divisions that are connected to prefectural and national professional associations. All public upper secondary school teachers of science are members of professional associations of science teachers organized according to the four discrete science subjects. Examples are the Japan Society of Physics and Chemistry Education (*Nihon Rikagaku Kyokai*) and the Japan Association of Biology Education (*Nihon Seibutu-kyouikukai*). All associations hold annual conferences; implement workshops, seminars, and field trips; publish teaching materials; and commend distinguished teachers.

Some science teachers join academic research associations as individuals. These associations include the Society of Japan Science Teaching (*Nihon Rika-kyoiku Gakkai*), Japan Society of Science Education (*Nihon Kagaku-kyoiku Gakkai*), the Physics Education Society of Japan, the Chemical Society of Japan (Education and Promotion Division), the Society of Biological Science Education of Japan, and the Japan Society of Earth Science Education. Other science teachers join non-profit organizations that offer practical science lesson study (*jugyo-kenkyu*) or develop teaching materials. Among these organizations are the Sony Science Teachers Association, the Association of Science Education (*Kagaku-kyoikukenkyukyogi-kai*), the Research Society for Hypothesis- Verification-through-Experimentation Learning (*Kasetsu-Jikken Jugyo Kenkyu-kai*), and the Teachers' Organization for Skill Sharing (TOSS).

Issues in Upper Secondary Science Teacher Quality

Teacher Supply

Entry to school teaching as a career is generally conducted on a competitive examination basis. The examinations are set and administered by prefectural boards of education and private institutions. In 2002, the average "competition rates" (number of candidates divided by number appointed) for the teacher appointment examinations for public lower and upper secondary schools were 11.8 and 13.9. However, in 2012 the rates were down to 7.5 and 7.7. Those candidates who fail to qualify for entry may still be appointed as part-time personnel, and many such people undertake the examination annually until they pass.

In recent years, Japan has been facing an ageing population and low birth rate, a demographic pattern that has naturally affected the age structure of the teaching force. According to MEXT's School Teachers Statistical Survey (2014a), the proportion of teachers aged 50 years or more rose from 23 percent in 2004 to 37 percent in 2013 at the lower secondary level, and from 31 percent to 40 percent at the upper secondary school over the same period. At the same time, the proportion of teachers below age 30 showed a slight upward trend from 9 percent in 2004 to 14 percent in 2013 at the lower secondary school, and from 9 percent to 11 percent at the upper secondary level. The average age of secondary school teachers has increased—from 42.9 years in 2004 to 43.9 years in 2013 at lower secondary school, and from 44.3 years to 45.3 years at upper secondary school. In 2012, the number of retirees and teachers leaving the profession was 9,500 for lower secondary schools and 10,500 for upper secondary schools. Most of the teachers leaving the profession are the younger and middle-aged teachers. In 2012, the number of teachers hired in lower secondary schools was 11,600. The number employed at the upper secondary level was 10,100.

The latest revisions (i.e., 2008 and 2009) of the courses of study for lower and upper secondary education require all students to take a wider range of subjects than was previously the case and to study science. The change regarding science has meant more science study hours and a commensurate need for more science teachers. However, many science expert teachers who served in large numbers during the decades of rapid economic growth in Japan have now retired. Upper secondary school physics is particularly affected, as comparatively few science graduates present physics as their major. However, the situation is likely to rectify itself as demographic changes see a decline in the numbers of students and schools over the next 15 or so years.

Professional Education and Training Issues

The imbalanced age structure of the Japanese teaching force militates against the implementation of progressive teaching practices such as inquiry-based science education. A 2001 report from the National Teacher Training Universities recommended that those universities jointly develop a model core curriculum for teacher training with a view to filling the gaps between pedagogical and content knowledge (leading to improved PCK) and thus strengthening the links between theory and practice for their teacher candidates. Many teacher training universities have accordingly tried to reform their undergraduate education programs. The Japan Association of Universities of Education, for example, established a model core-curriculum research project group that examined the curriculum frameworks for elementary and lower secondary teacher training in the member universities. The association did not consider upper secondary teacher training in its own right, firstly because the teaching and professional subjects required for teacher certification for upper secondary schools strongly overlap with those for lower secondary schools and secondly because the universities do not distinguish between the two.

In 2008, MEXT implemented a new type of graduate school—professional schools for teacher education. The aim of these schools is to lift teachers' teaching ability to Master's level through a program that includes practice teaching and action research; there is no thesis component. MEXT plans to extend the professional schools for teacher education model to all national teacher training universities over the coming years in preference to the academic research type graduate schools of education. Many universities are actively engaged in changing their graduate school curriculum to a more practical one because of their commitment to producing high-quality practitioners.

Japan has neither national nor prefectural standards for assessing candidates for admission to university preservice teacher training programs and teacher certification. Prefectural teacher appointment examinations involve pen-and-paper tests on pedagogical knowledge, content knowledge, and PCK. These themes are also targeted by a group discussion skills test and personal and group interviews. MEXT sometimes uses pass rates in these teacher appointment examinations at the local prefecture level as proxy measures of the national teacher training universities' educational outcomes.

Issues at the Chalkface

Japan considers observation, experimentation, and other scientific activities to be very important elements of science teaching and has long written these into the subject objectives of the courses of study for primary through to upper secondary school. MEXT has been improving science education facilities, including the equipment used at schools for observation and experiments, in a systematic manner in accordance with the Science Education Promotion Act 1953. Each year, the ministry also subsidizes prefectures with half of the required amount for science education equipment maintenance costs. In 2009, before full implementation of the current courses of study, MEXT provided, as part of its commitment to improving the educational environment for science, an additional 40 billion yen for new facilities for school science.

In the compulsory schools, science teaching is carried out as inquiry-based learning in science laboratories or homerooms. During problem solving (*mondai-kaiketsu*), which is commonly equated with inquiry-based learning, students are expected to undertake learning tasks based on familiar natural items and events, verify through experiments and group discussions their expectations or hypotheses as based on their prior knowledge, and reconstruct their understandings more scientifically. Many teachers conduct numerous demonstrations and facilitate student experiments, setting aside science textbooks as they do so in order to promote proactive learning and assess student learning through criterion-referenced evaluation. Teachers are required to evaluate four assessment domains for every subject area at every level. The domains are interest, motivation, and attitudes; thinking, judgment, and expression; technical skill; and knowledge and understanding. Student experiments are regarded as good opportunities to assess students' authentic abilities relative to science. Science teachers generally strongly encourage students' abilities to plan, hypothesize, and make discoveries. On the whole, teachers prefer that their students do not first learn the science and then apply it deductively to experimental outcomes. Their philosophy of science education accordingly remains primarily empiricist or positivist because they want their students to find and understand scientific knowledge through experimentation.

Science teachers in the upper secondary schools usually use science laboratories for their lessons, and most of them have at least one science laboratory assistant in attendance. However, science teachers at this level tend not to implement student experiments as much as their colleagues at lower school levels do, preferring instead to rely on lectures and textbooks. In 2008, the Japan Science and Technology Agency and the National Institute for Educational Policy Research carried out the Upper Secondary School Science Teachers Survey (Center for Promotion of Science Education, Japan Science and Technology Agency 2010). According to the report, 70 percent of upper secondary science teachers said they put a lot of effort into science lessons, 20 to 40 percent of the teachers (depending on which discrete science they taught) said that they gave good coverage to the latest developments in science and technology, and 30 to 50 percent of them said that they forged strong links between science and everyday life. These percentages tended to be higher than the percentages in corresponding surveys conducted with elementary and lower secondary school teachers. However, less than 10 percent of the upper secondary science teachers who participated in the survey said that they encouraged their students to come up with experimental designs themselves or to present their ideas to others. A similar small number of teachers claimed to have their students conduct experiments more than once a week; this number was considerably lower than the numbers given in the commensurate primary and lower secondary school surveys. In addition, 60 to 80 percent of the upper secondary teachers answered "fewer than three hours per year" when asked to report how often their students engaged in science inquiry activities. When asked to give reasons why so little time was given over to students doing experiments, teachers identified lack of lesson time, the competing need for university entrance examination preparation time, and lack of facilities and equipment. The last of these factors relates to the tight budgetary situation many schools continue to experience with regard to science facilities and equipment. For instance, the national budget for science consumables is presently only 510 yen per student per year.

Upper secondary teachers are under pressure to impart the content knowledge that students need to pass university entrance examinations, which is what students expect from science education at that level. Many upper secondary school teachers are unfamiliar with promoting students' proactive learning and with assessing students' learning outcomes via criterion-referenced evaluation.

Since 2002, MEXT has been identifying upper secondary schools that place strong emphasis on science and mathematics education. Its purpose in this regard is to help these schools, termed Super Science High Schools (SSHs), foster excellent human resource development in science and technology through the implementation of enriched curricula, teaching methods and materials, and assessment methods in collaboration with universities and research institutes. The Japan Science and Technology Agency lends financial support to these special SSH activities. At present, about 200 upper secondary schools have the designation of SSH (Department fort Promotion of Science Education, Japan Science and Technology Agency 2014). A 2008 survey conducted by the Japan Science and Technology Agency and the National Institute for Educational Policy Research showed that more than 80 percent of SSHs had dedicated classroom hours for students to present their scientific research and express their ideas in English. Most SSHs had also incorporated lectures by scientists and engineers into their science programs.

Continuing Professional Development Issues

Prefectural boards of education provide all newly hired public school teachers with induction training in the form of a one-year program. The program involves a combination of in-school and off-school training. In-school training is carried out over 10 or more hours per week (300 hours minimum for a year) and is guided by the mentor teacher at the school. New teachers learn more about teaching techniques and the type of background pedagogical knowledge that teachers typically need by observing expert teachers' lessons or performing some lessons themselves while being observed by the expert teachers and then engaging in a lesson-study discussion with them. Off-school training is carried out across 25 or more days per year at prefectural educational training centers. All new teachers attend lectures and undertake exercises relating to subject teaching, school administration, student guidance, and teachers' mental health. They also have access to other kinds of training such as fieldwork and work experience in different job fields. Schools can use some of their budgetary allowance to hire stand-ins for new teachers receiving induction training.

Although teachers are required by law to participate in inservice training, attendance is not always actively pursued because it is difficult for teachers to secure the various allowances, time, and replacement teachers necessary for their participation. This situation needs to be contextualized with reference to the long hours and heavy duties that teachers experience.

About 70 percent of SSH science teachers meet with university academics or research institute experts at least once a year, compared with about 30 percent of science teachers in general. Some of the schools that have held the SSH designation for ten or more years and thus become "core SSHs" are now required to share the knowledge and skills of inquiry-based science teaching they have acquired from their practice with science teachers in neighboring schools. Take, for example, Ichinomiya High School, one of the top-ranking public general academic upper secondary schools inside Aichi prefecture. It was appointed as an SSH in 2003 and became a core SSH in 2013. It now provides induction and inservice training workshops for science teachers of Aichi prefecture in relation to curriculum development, lesson design, and teaching strategies for the new subject SPS. Workshops also focus on science exploration activities conducted by SSH senior (head of department) expert science teachers. The aim of this component of the workshops is to develop science teaching ability, particularly among young science teachers.

Trends and Developments in Upper Secondary Science Teacher Quality

It is often said that current upper secondary school education, university education, and the university admissions system in Japan have tended to require rote learning and reproduction of knowledge rather than authentic academic achievement, which includes the ability to think, judge, and express, as well as attitudes conducive to cooperating with a variety of people while maintaining an independent mind. A report recently released by the Central Education Council (2014) recommended that Japan's courses of study should be revised dramatically, especially in terms of specifying the target physical and intellectual abilities and ways of learning, such as active learning. A new achievement test, designed to improve student learning, is to be introduced for upper secondary school seniors in 2020, while a new common university admission test, aimed at evaluating in particular "thinking, judgment, and expression" will be introduced in 2021.

MEXT is intent on introducing criterion-referenced evaluation for upper secondary schools such as has already been rooted in the practice of compulsory-level schooling, and it has designated some public schools and educational and research institutions as agents to promote the development of evaluation techniques for a variety of learning outcomes. Ichinomiya Minami High School, an upper-middle-ranked public general academic upper secondary school inside Aichi prefecture, is one of these. It serves schools in the Aichi prefecture by helping teachers prepare lesson plans and evaluation guides and then following up with lesson studies of inquiry-based lessons. It subsequently helps schools revise those lesson plans and evaluation sheets in collaboration with other teachers, research supervisors of Aichi Prefectural Education Center, and a science education professor from Aichi University of Education.

The model provides inservice training based on action research conducted in group settings, and it is one that has yet to be widely used in Japanese schools. It promotes effective communication about teaching methods among teachers of different generations and between schools, universities, and educational training centers. Although the majority of upper secondary school science teachers are general university graduates with only the minimum number of credits required for teaching and have not acquired sufficient knowledge of students' cognitive processes and their assessment, the educational theories and research findings provided by the outside experts will fill the gap and support them in developing improved practices for the future.

References

Center for Promotion of Science Education, Japan Science and Technology Agency (JST). 2010. *Heisei 20nendo Kotougakkou Rika-Kyoin Jittai-Tyousa Houkokusho* [FY2008 Upper Secondary School Science Teachers Survey]. Saitama: Author, http://www.jst.go.jp/cpse/risushien/highschool/cpse_report_009.pdf.

Central Education Council. 2014. *Shotou-Tyutou Kyoiku Bunkakai Kotougakkou Kyoiku Bukai Shingi Matome: Kotoukyoiku-no Shitsu-no Kakuho/ Kojou-ni-mukete* [Summary Report of the Subcommittee of Upper Secondary School Education for the Committee of Elementary and Secondary Education: Towards ensuring and improving the quality of upper secondary school education]. Tokyo: Author, http://www.mext.go.jp/component/b_menu/shingi/toushin/__icsFiles/afieldfile /2014/07/25/1349740_1.pdf.

Department for Promotion of Science Education, Japan Science, and Technology Agency. 2014. *2014–2015 Super Science High School*. Saitama: Author, http://ssh.jst .go.jp/ssh/public/pdf/h26 ssh.pdf.

Ministry of Education, Culture, Sports, Science, and Technology (MEXT). 2009a. *Kotougakkou Gakushu Shidou Yoryou* [Courses of study for upper secondary schools]. Kyoto: Higashiyama Shobo.

———. 2009b. *Koutougakkou Gakushu Shidou Youryou Kaisetsu: Rika Hen/Risu Hen* [Teaching guide for courses of study for upper secondary schools: In science /science and mathematics]. Tokyo: Jikkyo Shuppan.

———. 2014a. *Gakkoukyouin Tokei-tyosa: Heisei 25nendo* [Statistical survey of school teachers FY2013]. Tokyo: Author, http://www.mext.go.jp/b_menu/toukei /chousa01/kyouin/kekka/k_detail/134561444.htm.

——. 2014b. *Heisei 25nendo Koritsu-kotougakkou-ni-okeruKyoikukatei-no Hensei / JissiJokyo-tyosa-no Kekka-ni-tsuite* [Results of the survey on the organization and implementation of the public upper secondary school curriculum in FY2013]. Tokyo, Japan: author, http://www.mext.go.jp/a_menu/shotou/new-cs/__icsFiles /afieldfile/2014/05/27/1342498_02.pdf.

Ministry of Education, Science, and Culture. 1980. *Japan's Modern Educational System: A History of the First Hundred Years.* Tokyo: Author, http://www.mext. go.jp/b _menu/hakusho/html/others/detail/1317220.htm.

Further Reading

Council on Modalities of the National Teacher-Training Universities. 2001. *Kongo-no Kokuritu-no Kyoinyosei-keiDaigakugakubu-no Arikata-ni-tsuite* [The future direction of undergraduate education systems for national teacher-training universities]. Tokyo: Author, http://www.mext.go.jp/b_menu/shingi/chousa/koutou /005/toushin/011101.htm.

Cummings, William K. and Philip G. Altbach. 1997. *The Challenge of Eastern Asian Education: Implications for America.* Albany, NY: State University of New York Press.

Isozaki, Takako and Tetsuo Isozaki. 2011. "Why Do Teachers as a Profession Engage in Lesson Study as an Essential Part of Their Continuing Professional Development in Japan?" *International Journal of Curriculum Development and Practice* 13 (1): 31–40.

Isozaki, Tetsuo. 2014. "Lesson Study Research and Practice in Science Classrooms." In *Encyclopedia of Science Education*, edited by Richard Gunstone. Dordrecht: Springer. doi: 10.1007/978-94-007-6165-0_412-1

Ogawa, Masataka. 2014. "Occupational Culture as a Means of Professional Development for Preservice Science Teachers in Japan." In *Innovations in Science Teacher Education in the Asia Pacific*, edited by Chen-Yung Lin and Ru-Jer Wang, 61–80. Bingley: Emerald Group Publishing Limited.

Stigler, James W. and James Hiebert. 1999. *The Teaching Gap: Best Ideas from the World's Teachers for Improving Education in the Classroom.* New York: The Free Press.

Tucker, Marc S. and Betsy Brown Ruzzi. 2012. "Japan: Perennial League Leader." In *Surpassing Shanghai: An Agenda for American Education Built on the World's Leading Systems*, edited by Marc S. Tucker, 79–111. Cambridge, MA: Harvard Education Press.

7

New Zealand

Peter Rawlins and Carrol Walkley

Teaching Science in New Zealand

Backdrop

New Zealand is a country of 4.4 million people situated in the southern part of the Pacific Ocean. Within the country, 15 percent of the population identify themselves as the indigenous Māori, 69 percent as New Zealand European, 9 percent as Asian, and 7 percent as Pacific Islanders (Pasifika). Education is provided free for New Zealand citizens and permanent residents and is compulsory for students between the ages of 6 to 16, although most students start school at age 5. In the majority of cases schooling is within the government-run education system, but there are a small number of integrated schools and private schools, and some students receive exemptions to be home schooled. Prior to beginning their compulsory schooling, children can attend a range of early childhood provisions; almost 96 percent of children do so.

Schools within the compulsory school sector are organized in different ways. Commonly, primary schools cater for students from Year 1 to Year 8 (full primary), or Year 1 to Year 6 (contributing schools) followed by intermediate schools, which cover Years 7 and 8. A small number of middle schools cater for students from Year 7 to Year 10. Although most secondary schools cover Years 9 to 13, some accept students from Year 7 through to Year 13. In some rural locations, "area schools" cover the full range from Years 1 to 13. New Zealand also has a well-developed correspondence school catering primarily for remotely located students as well as students who are unable to attend a face-to-face school because of special circumstances. There are no barriers to transition from one school type to another.

In order to more fully appreciate the nature of current senior secondary science education, it is helpful to briefly look at the history of the development of the school sector in New Zealand. The 1877 Education Act set up a nationwide system of free, secular primary schools for all children from age 7 to 14. It was hoped that the new primary schools would provide a more equitable universal public system that would not only give everyone a chance to succeed but also aid social cohesion and, by providing the work ethic and skills the country needed, improve economic productivity (Bull et al. 2010).

While primary education was seen to be schooling for all—providing the basics of reading, writing, and arithmetic—secondary education was deemed only necessary for those preparing for university and the professions. This division was significant with respect to curriculum development and particularly with respect to science education, as will be discussed later. Additionally, while primary schooling was free, secondary schools charged fees until 1914, effectively restricting access to those families who could afford it. A significant increase in the number of students attending secondary school in the 1920s resulted in concerns that the traditional academically oriented curriculum would not meet the needs of the new cohort of students. Experiments with "technical schools" offering more practical and relevant curricula for the working class were not successful.

Eventually, in 1944, the Thomas Report (Department of Education 1944) argued for a common core curriculum drawing on the traditional academic subjects as well as the more practical subjects. This recommendation saw the introduction of the subject "general science" designed to provide both practical science study as part of citizenship for all as well as pre-professional education for ongoing science study. Despite the best intentions of the Thomas Report, schools continued to allocate students to classes based on their general academic abilities and often offered different versions of the "core" curriculum to different classes, thus perpetuating the academic/practical split.

Although New Zealand has no official requirement for students to show satisfactory progress at a given school level before going on to study at higher levels, schools do develop and apply informal rules governing student progression to higher-level study. Transition to university is controlled by the awarding of university entrance based on required levels of achievement in the National Certificate of Educational Achievement (NCEA) Level 3 (Year 13) as well as numeracy requirements at Level 1 (Year 11) and literacy requirements at Level 2 (Year 12).

School Science

At present, New Zealand has a compulsory school curriculum from Years 1 to 10 that includes eight essential learning areas, including science. From the earliest levels, the science learning area is divided into four contextual strands—the living world, planet Earth and beyond, the physical world, and the material world, as well as an overarching skills strand called the nature of science. Until students reach secondary school, science is generally taught by non-specialist teachers. It is common for science at these levels to be investigative and exploratory in nature and to be integrated into a wider unit of work. In many instances, students often do not realize that they are "doing science."

For students starting high school, science at the Years 9 and 10 levels is a stand-alone subject that integrates the four strands of the science curriculum. The three-year formal school exit qualification system—NCEA—starts for most students in Year 11. Although science is not compulsory in Year 11, approximately 90 percent of students take science, with approximately 15 percent taking one or more of chemistry, biology, and/or physics as an additional subject.

Science at the senior secondary school level is still largely taught as the three traditional subjects of physics, biology, and chemistry. Although officially the science curriculum has a duality of purpose, in reality the focus of secondary science centers on pre-professional education.

New Zealand has traditionally done reasonably well in international measures of student achievement. In the Programme for International Student Assessment (PISA) conducted by the OECD, New Zealand students have, on average, been placed well above the OECD average. Their placement on the Trends in Mathematics and Science Study (TIMSS), conducted by the International Association for the Evaluation of Educational Achievement, has been either just above or below the international median. In particular, the proportion of students considered to be high achievers on both studies is greater than many similarly positioned OECD countries.

Of concern, however, is the widespread achievement range of New Zealand students and particularly the disproportionate number of Māori and Pasifika students below the accepted minimum achievement benchmarks. In recent cycles of these international measures of achievement, New Zealand's place has softened. Increasingly, however, questions have been raised about the validity and reliability of these measures as true indicators of student achievement. New Zealand's recently revised National Monitoring Study of Student Achievement may give a more accurate picture of students' abilities in science and of the areas that can be developed to improve student outcomes.

Teacher Academic and Professional Education and Training

The most common pathway to training as a secondary teacher is to complete a minimum of a three-year content-oriented degree (e.g., a Bachelor of Science in biology) followed by a one-year initial teacher education (ITE) qualification. Students wanting to teach science generally need to have studied the relevant science subjects to at least second-year university level if they want to teach junior science or to third-year level if they want to teach senior science subjects.

Students apply for entry to a one-year graduate diploma ITE course at one of the country's nine New Zealand Teachers' Council-accredited providers. In addition to candidates having the necessary teaching subject content knowledge, selection is based on suitability to be a teacher. The one-year courses are a mix of theoretical and practical aspects of becoming a teacher (with the focus on pedagogical knowledge and pedagogical content knowledge) and also include a minimum of 14 weeks' practicum experience in schools.

Having gained a teaching qualification, teachers then apply for provisional registration and teaching positions. Graduates have a minimum of two years as a provisionally registered teacher, during which time they are under the supervision of the school in which they work. In their first year of provisional registration, teachers are given only a maximum of a 0.8 teaching load with respect to contact hours; in their second year, a maximum of 0.9. During this time, schools are required to provide an induction and mentoring program, and those teachers are required to gather evidence of progress toward meeting New Zealand's registered teacher criteria (see below). Although New Zealand does not have a prescribed program of training, one criterion that teachers have to meet is engagement in continuing professional development (CPD). Teachers usually meet this obligation by accessing externally provided training in an area identified as needing development. This could be in classroom management or may be in subject knowledge, assessment, or pedagogy.

Continuing Professional Development

The New Zealand Teachers Council (NZTC), an autonomous Crown entity that serves as a professional and regulatory body for teachers, has established 12 criteria that teachers need to meet in order to be registered. One of these is "Demonstrate a commitment to ongoing professional learning and development of personal professional practice."

Professional development provision has historically been provided by the universities, which were awarded contracts funded by the Ministry of Education. The ministry would decide on an area to be developed and would then invite schools to send teachers to the one-day courses. This "one size fits all" approach had limited success and often did not meet the immediate needs of teachers. One-day workshops, moreover, have limited effect without follow-up action, and this has been a contributory factor in their drop in popularity.

More recently, the Ministry of Education has allocated funding for CPD to the schools, allowing them to determine their needs and then to "buy in" the advice they want. This development has placed the university providers in an uncertain position in the absence of direct ministry funding; many have had to make staff redundant. Many of the university-based advisors' positions no longer exist, and science teachers no longer have the university-based science advisor as a resource. The reduction in advisory positions has meant a loss of a considerable amount of expertise and many long-standing relationships built up over the years.

Some of the university-based advisory services became private companies as the universities focused on core business activities and shed some of their commercial operations in the changing environment. Independent providers also grew in number. These changes have resulted in more targeted CPD for schools, and the one-day workshops have largely been replaced with more sustained ongoing in-school coaching models.

The last two decades have seen a steady interest in teachers taking further study leading to higher qualifications. This advanced study is generally taken up by teachers holding middle management responsibilities and thus needing to build expertise in leadership and pedagogy covering school-wide issues and so is not specific to science teaching.

Science Teacher Professional Associations

A number of professional associations support science education in New Zealand. The overarching body is the New Zealand Association of Science Educators (NZASE), which acts as a central hub for science educators across a range of levels and is closely linked to its parent body, the Royal Society of New Zealand. NZASE has an extensive website and also publishes the *New Zealand Science Teacher*, a publication that provides "news, opinion, videos, and comprehensive articles (both academic and informative) relevant to science education in New Zealand." It is available free to NZASE members in both print and digital form (iPad and Android).

NZASE also has close links to a range of subject associations. These include the New Zealand Association of Primary Science Educators and a number of secondary school-focused associations (the Biology Educators' Association of New Zealand, Horticulture and Agriculture Teacher Association of New Zealand, Earth and Space Science Education of New Zealand, and the New Zealand Association for Environmental Education). Additionally, organizations such as the New Zealand Institute of Physics, the Institute of Professional Engineers New Zealand, and the New Zealand Institute of Chemistry cater for their members, including those involved in education across the sectors. NZASE also has a branch association to support science technicians in the education sector (Science Technicians' Association of New Zealand).

NZASE and the connected subject associations provide extensive support for senior secondary school science teachers, with that support including teacher professional learning, conferences, curricula, pedagogy, and assessment resources. Subject-based conferences in biology, chemistry, and physics take place every second year and alternate with the national science conference SciCon. These combine presentations on pedagogy, assessment, and updates from the Ministry of Education and the New Zealand Qualifications Authority (NZQA).

While nationally the subject-based associations and organizations remain relatively strong, at the branch level they have been in decline in recent years. The increased administrative workload of teachers, often caused by the NCEA, combined with the disestablishment of the university-based science advisors' positions, which teachers often relied on for leadership, have seen a decrease in activity in many regions. The decline in regional subject associations combined with the reduction in the number of one-day CPD workshops has resulted in reduced opportunities for teachers to network with other schools.

Issues in Upper Secondary Science Teacher Quality

Teacher Supply

In general, the Ministry of Education has not considered teacher supply and retention an issue in recent years. The ministry's latest indications are that retention has remained at a constant level and that the loss of teachers is consistent across the regions and school deciles (rankings by socioeconomic profile). The age profile of the teaching workforce shows that, between 2004 and 2008, the proportion of teachers between 55 and 70 years of age steadily increased each year from 19 percent to 24 percent (Engler 2008). The expectation that the "baby boomer" teachers in the older bracket would all suddenly retire, causing an acute shortage, does not appear to be happening. Reasons for this situation are many, but a commonly held view is that the global financial crisis led to many teachers losing retirement savings and therefore having to continue working.

In recent times, ITE providers have noted a decline in the numbers of people applying to enter their programs, a situation that is partly attributed to media coverage of the oversupply of teachers and the reduced number of teaching positions advertised. Given the long-term situation of only a small number of applicants qualified in physics and chemistry, any decline in overall numbers is worrying. With fewer applicants, universities—as economic entities—face the challenge of ensuring their pre-selection process is robust and successful applicants are "up to standard."

Despite the lack of concern over the supply and retention of teachers, staffing some subject areas remains difficult; physics and chemistry have long fallen into this category. The Ministry of Education provides incentives through TeachNZ scholarships for ITE students in difficult-to-staff subjects. However, with the exception of Te Reo Māori (Māori language), these scholarships have recently been discontinued, although the reason for their demise appears to be primarily financially based.

Another issue facing schools is that of teachers being required to teach in a subject they have neither majored in nor been trained in during their ITE. In New Zealand, senior and junior secondary schooling resides in a single institution, which means that, in general, specialist science teachers working in a secondary school teach a range of classes from Year 9 to Year 13. The well-documented shortage of physics teachers, for example, can present the necessity for a chemistry or biology graduate to teach the subject. Additionally, on occasion, some junior science classes may be taught by, for example, physical education graduates. This scenario can occur when schools have extra capacity in some teaching areas and a shortage of qualified science teachers. While taking up this "slack" can lead to professional growth for the teachers concerned, it nonetheless depends on those teachers gaining sufficient content and pedagogical content knowledge to provide quality learning experiences for students rather than simply engage in transmission teaching.

Professional Education and Training Issues

A number of preservice ITE issues merit consideration. While these issues are not specifically or uniquely related to training science teachers, they are relevant to bringing quality teachers into the profession, including quality science teachers. Recently, various educational stakeholders have expressed concern about perceived political pressure placed on the design and provision of

ITE. For example, in 2013, the Ministry of Education funded pilot schemes in two universities to provide a Master of Teaching qualification to commence in 2014. The ministry argued that this initiative was "consistent with international moves to lift the quality of teaching and ensure that teachers have the competencies to work effectively in twenty-first century learning environments" (Ministry of Education 2014). While this initiative has been generally welcomed by ITE providers and the wider teaching community, some are concerned that it signals a shift toward more direct government involvement in the design and provision of ITE programs. Concerns relate not to the provision of postgraduate ITE per se, but rather the manner by which the government, via the Ministry of Education, is deciding what these programs will look like and who will deliver them. The government's move toward postgraduate ITE is being watched with particular interest, given that the government has already foiled attempts by at least two universities to provide postgraduate ITE programs that the universities had designed themselves.

Prior to the government initiative, providers designed their ITE programs and then sought accreditation for them from the New Zealand Teachers Council. Essentially, any provider who met the council's broad accreditation guidelines could run an ITE program. Under the new government initiative, only those universities whose proposed programs meet the government's "vision" and specific guidelines for ITE have been approved and funded.

Numbers in Master's programs are limited, with a higher than usual grade point average for entry, currently set at a B+ average. Moreover, these programs are not currently offered as extramural distance initiatives, meaning that student teachers must be in a position to move to the provider's city. Concerns have been raised that these conditions will restrict access to the detriment of potentially good teachers.

Included in the government's requirements is a closer partnership with schools. Trainee teachers are expected to spend significant amounts of time each week in schools, with teachers and teacher educators co-developing and co-teaching the ITE program. On the surface, this requirement seems eminently sensible, but concerns have been raised over how this partnership will evolve and whether existing teachers have the time, expertise, or qualifications to take on this new role of delivering the ITE curriculum directly. Not all good teachers make good teacher educators. During 2015, the pilot was widened to include more providers. The success of these courses will be assessed, and the results of that assessment awaited with interest.

Other recent initiatives have also raised concern about increasing government control of the teaching sector. In 2013, the Minister of Education announced that the New Zealand Teachers Council would be replaced by a new body called The Education Council of Aotearoa New Zealand. More than a mere name change, this move signaled a significant change in the constitution of the governance body responsible for registration and regulation of teachers. The New Zealand Teachers Council governance body consisted of 11 members: 4 Minister of Education appointees, 4 elected teacher representatives, and 3 members appointed from the sector unions and the New Zealand

School Trustees Association. The new Education Council governance body will consist of 9 members appointed directly by the minister for education. Considerable concern has been expressed as to whether this new body will be sufficiently free of government influence.

A further problem that has arisen in recent years is the provision of quality associate teachers to work with student teachers while they are on practicum. ITE programs rely on schools hosting student teachers as they complete their required 14 weeks of practicum in schools. During their time on practicum, student teachers are mentored by associate teachers, who receive a modest payment. The selection and approval of associate teachers is solely at the discretion of the host school, and the possibility exists for variability in the level of mentoring provided. Furthermore, increases in teachers' workload have resulted in a reduction in the number of teachers prepared to act as associate teachers, and not all teachers see being involved in student teacher training as an obligation or, indeed, a privilege. As such, it is becoming increasingly difficult to find host schools for student teachers, particularly in less popular subjects. This has been less of an issue for science student teachers because science teachers are generally sought after and schools use the hosting on practicum as an opportunity to recruit new staff. Nonetheless, difficulty finding quality associate teachers remains a concern given the government's move toward closer partnerships in the new Master of Teaching qualifications discussed earlier.

Issues at the Chalkface

In 2010, the New Zealand government introduced national standards—a mandatory testing and reporting system for literacy and numeracy—into primary schools. Evidence suggests that, as many commentators anticipated, the introduction of these standards is leading to an increasing emphasis on literacy and numeracy in the primary classroom at the expense of other areas of the curriculum, including science (Thrupp and White 2013). Professional development over the last five years has also centered on literacy, numeracy, and assessment practices rather than on broader curriculum areas. A lack of emphasis on science, coupled with primary teachers' traditionally low levels of science expertise and low confidence in teaching science (Gluckman 2012), raise concerns for the future achievement levels of primary students entering secondary school.

A significant challenge to teachers over recent years relates to the government's implementation of achievement standards (assessment units) and, in particular, internal assessment of practical work. While teachers argued that the implementation of the NCEA was under-resourced and left to a large extent to the schools, the Ministry of Education, acting on government directive, opined that this process gave ownership of the assessments to schools. One thing that is clear is that this development has created an even greater increase in workload for teachers.

During the early years of NCEA implementation, sourcing sufficient laboratory equipment was considered problematic. However, resourcing, in general, now appears to be of minor concern to most school science department. This reduction in concern is primarily due to an increased recognition of the importance of sufficient equipment to allow students to perform well in internal assessments, the results of which are publicly available and used as a proxy measure for school quality. A positive aspect of this situation is that schools are now obliged to keep laboratories fully equipped with equipment in good repair. A progressive refining of assessment tasks has also reduced the resourcing issue. For example, earlier internal assessment provided for gravimetric and colorimetric determinations in chemistry, but the expense of the equipment needed to run these assessments was prohibitive. Skills requiring use of this equipment are no longer assessed.

Most laboratories are equipped with a full range of equipment, and nearly all classes have data projectors and access to the Internet. Predictions of computer simulations taking the place of carrying out experiments has not occurred to any great extent, except in the case of dissections in biology classes. On the whole, many laboratory skills are assessed and need to be practiced with actual equipment rather than through simulation.

Unlike in the past, when it was difficult to attract suitable technicians, working as a science technician in a secondary school now appeals to a wider range of people. Schools generally have a skilled technician helping to run the science department and maintain and prepare equipment. Their help is important in running the high-stakes assessment of practical skills for the NCEA. Larger urban schools may have a hundred students taking an assessment involving titration or microscopy skills in an examination period. Teachers rarely have the time to organize and set up the equipment on their own.

Significant in terms of how teachers teach science are their views on the purpose of science education and the nature of science in the curriculum. In his 2011 report on the current state of science education in New Zealand, the Prime Minister's Chief Science Advisor Sir Peter Gluckman identified two broad purposes for science education at school (Gluckman 2011). The first was "pre-professional education." This, Gluckman argued, is what most parents think of as the traditional role of science education—preparing students to enter tertiary study by imparting the necessary knowledge of physics, biology, and chemistry. This view is generally consistent with the traditional academic focus of the science curriculum prior to 1944.

The second purpose Gluckman identified was "citizen-focused objectives," which involves preparing students for the complex role that science will play in their adult lives. Gluckman drew on the work of Bull et al. (2010) to break this purpose down into three broad areas:

- A utilitarian purpose, where students gain a basic knowledge of how things work (e.g., the human body) sufficient for them to appreciate the technological, environmental, and biological world in which they live and work;
- A democratic/citizenship purpose, which equips students with the necessary scientific literacy to take an active part in science-related

discussions and debates (e.g., adding fluoride to water supplies, or exploring the causes and effects of climate change);

- A cultural/intellectual purpose, directed toward developing students' skills in scientific thinking as part of their general intellectual development and thus allowing them, for example, to examine the reliability of information, which is an increasingly important skill given today's easy access to information via the Internet.

One could argue that the pedagogies appropriate for teaching these citizen-focused objectives differ considerably in nature from many of the pedagogies commonly found in traditional science education at secondary school. The citizen-focused objectives are achieved through much more investigative and inquiry-based pedagogical approaches encouraging reasoning, scientific thinking, and debate related to issues relevant to students. While a modern pre-professional course should also be able to aspire to these approaches, and indeed the New Zealand Curriculum encourages this, realizing these objectives remains a challenge for senior secondary school education in New Zealand.

Another issue, one that relates to the purposes of science, is the struggle that schools have experienced over the last 60 years as they have grappled with the nature of science in the curriculum, the types of pedagogies that should be enacted, and the types of learning experiences that students should have. The ensuing dilemmas have resulted in two dominant philosophies. The first encompasses "knowledge-centered" approaches that view science as a body of objective content, facts, and principles to be "transmitted" to students. The second embraces "learner-centered" approaches that seek to build on students' naturally inquisitive nature and interest in the world around them. Despite official attempts to marry these two philosophies in official curriculum documents and in the redesign of the high-stakes exit qualification system, the learner-centered approach is more prevalent in primary school while the knowledge-centered approach dominates in secondary school. This focus on learning "content" in secondary school has been identified as a barrier to students wanting to continue to study science for its own sake (Bull et al. 2010). And despite the knowledge-centered focus in senior secondary school, universities still express concern that many of the students who apply to study professional science courses still lack the necessary content knowledge.

While the New Zealand Curriculum guides what—and more explicitly how—science is taught up to Year 10, the demands of NCEA assessment have a significant influence in the classrooms of senior-years teachers. Achievement standards are graded as "not achieved," "achieved," "achieved with merit," and "achieved with excellence." Each grade reflects a level of increasing cognitive difficulty. "Achieved" requires a description, "merit" an explanation, and "excellence" a comparison of two explanations. These requirements are informed by Biggs and Collis's (1982) Solo Taxonomy of Understanding, and there has been increasing interest in the practical application of this taxonomy in classrooms.

With the knowledge that students are to be assessed on their ability to give written explanation to show their understanding, teachers spend significant time on practicing this skill and giving formative feedback. The days of multiple-choice questions, one-word answers, one-sentence definitions, and labeling diagrams have passed into history. Today's science teachers, like all teachers, are teachers of literacy. Changes in assessment have been a factor in moving teachers from the traditional pedagogy of transmission teaching of content to more contemporary pedagogies such as group work, problem solving, formative assessment, and peer- and self-assessment.

Practical work too has changed. Many teachers used to consider it optional and would use it to incentivize students to complete the "boring" theory work. The importance of practical skills has been validated by their inclusion in the formal assessment system. No longer do students think of practical work as "just for fun." Mastering the skills is now important: the more accurately a student can carry out a titration, the better his or her grade will be; the more accurately a biology student can focus a microscope on a leaf epidermis and draw a diagram, the better his or her grade. Additionally, because group investigations in science are also now assessed, teachers are much more aware of the importance of providing students with opportunities to learn and practice while working together in groups.

Another of the challenges facing science teachers is moderation of the results of the internally assessed achievement standards. Schools are expected to have in place an internal moderation plan so that work from several classes is marked to the same standard. Typically, one teacher marks all student work for each assessment. However, if the cohort is large, the head of department (i.e., the teacher with overall responsibility for a subject-area domain in a school) check-marks a sample of marked assessments from each teacher for consistency.

As part of New Zealand's national moderation system, teachers are required to send a randomly selected sample of marked student work to the New Zealand Qualifications Authority (NZQA) for moderation. The sample has recently been increased to 10 percent of each compendium of assessments after doubts were raised about the effectiveness of the moderation using a smaller sample of eight scripts from each school. National moderators check the validity of the assessment task and the accuracy of application of the marking criteria. They then provide feedback designed to help the school improve the quality of future assessment. Recent publicity has suggested that the error rate is not declining and that teachers are not becoming more skilled in making accurate judgments (NZQA 2014).

As a result of the NZQA policy that allows students to be reassessed for internal achievement standards (ASs), the success rate is high. When the examination scripts for the externally assessed ASs in science are marked, the results fit a profile of expected performance that typically has 10 percent of the student cohort achieving "excellence," 20 percent "merit," 40 percent "achieved," and 30 percent "not achieved." The internally assessed ASs do not have this constraint, and not only is the pass rate high but the percentage of students being awarded a grade of excellence is also high. There is no matching of achievement between the internal and external grades on the basis that they assess different skills. This disconnect has led to some distrust of the results gained by internal assessment and a suspicion that some schools may accept a lower standard or present the easier assessment tasks. Recently, a restricted-entry university course published advice that it would select students on the basis of Level 3 NCEA results gained only in external examinations.

Continuing Professional Development Issues

In the days when senior biology, chemistry, and physics were all assessed by three-hour external end-of-year written examinations based on a course common to all schools, preparation for assessment was reasonably straightforward. However, the past decade has seen significant changes in the assessment of senior subjects in New Zealand secondary schools with the introduction of standards-based assessment for the National Certificate of Educational Achievement (NCEA). Since it was introduced, the qualification has had a considerable influence on the nature of the CPD on offer, while the associated internal assessment, particularly of practical skills, has impacted significantly on all science subjects.

Introduction of the NCEA meant that schools needed to decide which achievement standards they should select when developing their courses. Most schools offer courses in the senior sciences that are made up of 5 ASs, totaling about 20 credits (ASs carry different numbers of credits). The New Zealand Qualification Authority's regulations require that schools offer courses with at least one externally assessed and one internally assessed AS. Teachers are also expected to design assessment tasks and carry out moderation of the results.

All of the changes associated with implementation of the NCEA have placed considerable demands on teachers' time and effort. As part of attempts to ameliorate this situation, CPD has focused on solving these problems. A prime example from the last round of changes to the ASs completed in 2012 was the introduction into Level 3 chemistry of AS 91388: "Demonstrate understanding of spectroscopic data in chemistry." Teachers wanting to include this AS in their course needed to generate resources and assessment materials as well as revise their own knowledge of spectroscopy. This topic had previously not been taught at secondary level and so created a demand for CPD, but in meeting this reactive demand, the emphasis has been taken from proactively looking at improving pedagogy to a focus on assessment.

Trends and Developments in Upper Secondary Science Teacher Quality

The New Zealand government recently announced a number of initiatives to improve student achievement levels. Included in these initiatives is the creation of two new roles, developed in recognition of the expertise available in schools. The first of these roles, the "expert teacher," works not only within his or her own school but is also released to work with other schools within the local school community to lift practice and improve achievement. It is expected that 1,000 expert teachers will eventually be in place, each of whom will receive an additional NZ$20,000 per annum. However, these teachers will be active in only a limited variety of targeted subject areas, including science. The second role, the "lead teacher," is to be held by highly capable teachers with proven track records able to act as role models for teachers within their own schools as well as in other local schools. The anticipated 5,000 lead teachers will receive NZ$10,000 extra pay per annum.

The new Master of Teaching programs discussed earlier are due to produce their first graduates at the end of 2015. These programs are designed to improve the quality of teaching and equip teachers for a twenty-first-century learning environment. The inclusion of the B+ entry criterion should improve the overall academic quality of the prospective teachers. It remains to be seen whether this expected improvement will translate into improving the quality of teaching in the classroom.

At the time of writing, the existing conjoint and graduate ITE programs were still being run, but it was not clear whether this arrangement would continue or whether it would be phased out in favor of "approved" postgraduate courses. If this phasing out were to occur, then it would potentially restrict the design of and access to ITE programs.

The NZQA has placed a moratorium on changes to the NCEA for the next five years in acknowledgment of the stress these continual changes have caused and the need to consolidate existing practice. This "breather" should increase familiarity with assessment guidelines and give teachers confidence to administer the internally assessed ASs. The moratorium will also allow time for data to be collected on which to base future changes as well as time to examine existing pedagogy with a view to continual improvement and perhaps for teachers to regain some creativity in their classrooms.

References

Biggs, John B,. and Kevin F. Collis. 1982. *Evaluating the Quality of Learning: The SOLO Taxonomy.* New York: Academic Press.

Bull, Ally, Jane Gilbert, Helena Barwick, Rosemary Hipkins, and Robyn Baker. 2010. *Inspired by Science: A Paper Commissioned by the Royal Society of New Zealand and the Prime Minister's Chief Science Advisor.* Wellington: New Zealand Council for Educational Research.

Department of Education. 1944. *Report of the Committee on the Post-Primary School Curriculum* (the Thomas Report). Wellington: Government Printer.

Engler, Ralf. 2008. "Teacher Loss Rate Report 2007–2008 Update" on the Education Counts, http://www.educationcounts.govt.nz/publications/series/2263/31417/4.

Gluckman, Peter. 2011. *Looking Ahead: Science Education for the Twenty-First Century.* Auckland: Office of the Prime Minister's Science Advisory Committee, http://www.royalsociety.org.nz/media/Science-education-PDF-version.pdf.

———. 2012. *Science and New Zealand's Future: Reflections from the Transit of Venus Forum. A Report to the Prime Minister from Sir Peter Gluckman.* Auckland: Office of the Prime Minister's Science Advisory Committee, http://www.pmcsa.org.nz/wp -content/uploads/Transit-of-Venus-Forum-report.pdf.

Ministry of Education. 2014. "More About the New Exemplary Postgraduate ITE Programmes," Education.govt.nz http://www.minedu.govt.nz/theMinistry /EducationInitiatives/QualityOfITEProvision/PostGraduateITEProgrammes.aspx.

New Zealand Qualifications Authority (NZQA). 2014. *Annual Report on NCEA and New Zealand Scholarship: Data and Statistics* (2013). Wellington: Author, http:// s3.documentcloud.org/documents/1180463/ncea-annualreport-2013.pdf.

Thrupp, Martin, and Michelle White. 2013. *Research, Analysis and Insight into National Standards (RAINS) Project Final Report: National Standards and the Damage Done* (report commissioned by the New Zealand Educational Institute TeRuiRoa). Hamilton: Wilf Malcolm Institute of Educational Research, http://www.waikato .ac.nz/__data/assets/pdf_file/0010/179407/RAINS-Final-report_2013-11-22.pdf.

Further Reading

Bay, Jacquie L., Richard Meylan, Joanna Leaman, Sarah Gibbs, and Alan Beedle. 2010. *Engaging Young New Zealanders with Science: Priorities for Action in School Science Education. A Discussion Paper from the Office of the Prime Minister's Science Advisory Committee.* Auckland: Office of the Prime Minister's Science Advisory Committee, https://researchspace.auckland.ac.nz/handle/2292/14193.

Education Review Office. 2012. *Science in the New Zealand Curriculum: Years 5 to 8.* Wellington: Author, http://www.ero.govt.nz/National-Reports/Science-in-The-New -Zealand-Curriculum-Years-5-to-8-May-2012.

New Zealand Association of Science Educators (NZASE), website, 2015, http://nzase .org.nz/.

8

South Korea

Nam-Hwa Kang

Teaching Science in South Korea

Backdrop

Since the end of the Korean War in 1953, South Korea has achieved remarkable economic growth. World Bank statistics showed South Korea's GDP to be the fourteenth largest in the world in 2013. The country's per capita GNI of $110 in 1962 had risen to $33,500 by 2013. Given that South Korea has few natural resources at its disposal, the country's achievement can largely be attributed to its human resources, the driving force behind which is deemed to be education. High levels of value and high expectations with regard to education pervade Korean society.

Since the inception of public education, South Korea has had a 6-3-3 education system—that is, 6 years of elementary education, 3 years of lower secondary (middle school) education, and 3 years of upper secondary (high school) education. Of these 12 years, the first 9 are compulsory and fully funded by the government; students are charged nominal fees for the last 3 years. The fees do not deter student enrollment in upper secondary education. In 2014, educational statistics showed that 99.7 percent of middle school graduates transited to the upper secondary school system. As of 2014, South Korea had about 7 million students in primary and secondary schools (for a population of approximately 50 million).

Students usually attend elementary and secondary schools near their homes. Students proceed from elementary to middle schools as long as they have met the mandatory minimum attendance requirement. The transition from lower to upper secondary schools has become more complicated recently because of the creation of various types of high school. There are four such types: general, special-purpose, specialized, and autonomous. Special-purpose high schools (a total of 143 across the nation) include foreign language schools, science schools, international schools, music and art schools, schools for athletes, and schools for the gifted. Graduates from these schools are expected to move on to university education to continue studying their respective specialty areas. Specialized high schools (a total of 499 nationwide) include vocational schools in various fields such as technology, farming, business, and fishery. While most graduates from these schools are expected to enter the associated vocations, some move on to university education. Autonomous high schools (164 in total) are established by people and organizations wanting to pursue an educational vision/agenda that differs from the national one. These autonomous high schools are given more leeway in implementing the national curriculum. General high schools (1,520 across the nation) include those students who will move on to higher education and those who elect to finish their formal education with only a high school diploma. Thus, general high schools have students with the widest range of abilities within a classroom. General high schools operate two tracks: humanities and sciences.

Processing for admission to non-general high schools begins near the end of middle schooling. Processing of applications for general high schools follows. Students can apply for any non-general high schools in the nation, and these schools have their own selection processes, including their own examinations. General high schools have common application processes within a city or province, which typically take into account grade point average, teacher observation records, and (in some provinces) a qualifying examination. Students are assigned to a school near their homes. Should a student fail to enter a general high school, which is a very rare occurrence, he or she can nonetheless take a high school diploma examination at any time.

High schools exhibit a range of tuition scales. General high schools and public autonomous high schools charge about US$1,200 per year (just over 3 percent of South Korea's average household income in 2014); low-income families can apply for scholarships. Tuition is almost completely subsidized for students enrolled in specialized high schools. Special-purpose high schools and private autonomous high schools charge tuition fees that are up to three and a half times more than the general high school fees.

Historically, Koreans have seen education as the main instrument of social mobility, and policies in modern South Korea have thus far enabled those with higher education to earn higher pay than the less educated, fueling people's expectations of education. Graduates of prestigious universities are especially advantaged in this regard. Government policy encouraging parents to be passionate about their children's education has created a high demand for better education and spurred the growth of South Korea's education sector.

The high expectations for education are evident in the continuing high rates of high school and higher education completion. According to the findings of the OECD Teaching and Learning International Study (TALIS; OECD 2014), South Korea had, as of 2012, the highest high school completion rate (98 percent) of all 44 participating countries and the highest (cross-nationally) higher education completion rate (66 percent) among people in the age group encompassing 25- to 34-year-olds. The OECD findings also showed the evident connection between education and securing employment for men. The employment rate at the time was 84 percent for those with only a high school diploma but 90 percent for those with a degree. Furthermore, those with a degree were earning 60 percent more than those with only a high school diploma (OECD 2014).

School Science

All South Korean schools, whether private or public, are required to follow the national curriculum. This curriculum has undergone regular change; the current one was developed in 2009. Within the national curriculum, science is offered as a subject beginning in Grade 3. Primary schools (Grades 3 to 6) are required to offer three class periods of science per week, exactly half of the class hours assigned for Korean language. Mathematics is required for four class periods per week at this level. Because primary schools use the same national science textbooks, primary school science education is uniform across the nation.

For the three years of lower secondary (middle) schooling, science classes are typically offered for three class periods weekly in Grade 7 and four class periods in Grades 8 and 9, the same as for mathematics. Textbooks for the middle-school level onward are published by private companies, although the books have to be approved for alignment with the national curriculum. In general, students in primary and middle schools are not tracked on the basis of ability, which means students learn the same national curriculum in mixed-ability classes.

From primary through middle school, science is taught as "general science," while at upper secondary level (high school) it is differentiated into disciplines (physics, biology, chemistry, and Earth science). High school science requirements differ across the different types of school. In order to graduate from general high school, students must complete a minimum of 15 units of science (one unit being 17 weeks of 50-minute lessons over a semester). In the humanities track, students take social studies and humanities elective courses and study science only to the minimum level required for graduation. Students in the science track take science elective courses above that minimum. Obviously, students in science high schools take many science courses, including advanced courses.

The national high school sciences curriculum includes general science for Grade 10 students, basic sciences (physics I, chemistry I, biology I, Earth science I) and upper-level sciences (physics II, chemistry II, biology II, Earth science II) for science-track students, and advanced sciences (e.g., advanced physics, advanced chemistry) for students in science-specialized schools. These are five-unit courses, although schools can decide to assign each subject one less or one more unit. Students combining science and humanities courses must gain 35 credit hours in order to graduate, and science subjects must account for at least 15 of these credits. In principle, students can choose their science or humanities subjects themselves. However, due to teacher supply, students are not completely free to choose electives. Humanities-track students typically take the minimum science requirement (general science and two basic sciences), while science-track students typically take 20 or more credits of science. A student may, for example, take general science, biology I, biology II, Earth science I, and Earth science II.

Given that the majority of high school graduates move on to higher education (71 percent in 2014), the high school curriculum is strongly affected by the demands of the College Scholastic Aptitude Test (CSAT). It is commonly said that high school education is geared to students successfully passing this university entrance examination. The CSAT is offered once a year in the second week of November. About 65,000 students sat it in November 2014. Of this number, about 20 percent were retakers—students who had waited another year or so to sit the examination again with the aim of achieving better scores. Some students take the CSAT several times in order to get high enough scores to enter the university or college of their choice ("Test-Taking in South Korea" 2013).

Students sitting the CSAT are currently required to take mathematics, Korean, and English tests, as well as two other tests in either social studies or sciences. Humanities-track students usually take two humanities or social science tests, while science-track students take two science subject tests. Because of the emphasis on CSAT and the two-track system in high school, a recent issue in educational policy has been that of high school graduates lacking a well-rounded education. Humanities-track students take the minimum number of science classes and hence lack scientific literacy, while science-track students lack literacy in social studies and humanities. In order to resolve this issue, the Ministry of Education is currently working on revising both the CSAT and the national curriculum accordingly. Effort is being made to create one eight-credit integrated science course and one eight-credit integrated humanities course, which all students studying for the CSAT must take and then be tested on.

Teacher Academic and Professional Education and Training

Teacher education in South Korea is nationally controlled in that the government stipulates certification requirements. Primary and secondary school teacher training programs are distinct. Individuals wanting to enter primary teacher training have to earn a teacher education degree at the Bachelor's level. At present, only 13 such programs are on offer in the nation, and they are all very similar due to the government's exact certification requirements. For graduation, 140 credits are required (one credit equals one hour per week per semester, i.e., 17 hours). Typically, 40 credits are for general education and 100 credits are for teacher certification courses. Among the 100 credits, about 60 are for subject-matter courses for all the subjects taught in elementary school classrooms and 20 additional credits are devoted to preservice teachers' specialty subjects such as science, mathematics, Korean language, physical science, and so on. The remaining 20 credits are for education foundation courses, including student teaching.

Candidates can take more than one route to becoming a secondary school teacher. The three most typical ways are as follows:

- Graduating from four-year secondary teacher programs at the Bachelor's level;
- Graduating with a Bachelor's degree in a school subject field such as physics or biology along with additional credits to meet the certification requirements; and
- Earning a Master's degree in subject-matter education along with completing the coursework for certification.

Secondary teacher education programs at the Bachelor's level usually require 140 credits for graduation, and the government requires students to take specific courses in line with certification requirements. For example, in order to attain physics teacher certification, a student must complete 21 credits of specific physics content courses and 9 credits of physics pedagogy courses. For chemistry teacher certification, a student has to obtain 20 credits of chemistry content courses and 9 credits of chemistry pedagogy courses. Biology teaching certification requires 29 credits of specific content courses and 9 credits of biology pedagogy courses.

Continuing Professional Development

Although the central government controls teaching certification and regulations regarding teacher inservice requirements, city and provincial education offices, known collectively as local educational agencies (LEAs), provide inservice training. There are 17 LEAs in South Korea. Teacher inservice training is also provided in the schools themselves and by teacher inservice centers. Of these centers, 4 are centrally controlled, 17 are governed by city and provincial education offices, 80 are attached to colleges of education at universities, 41 are online facilities, and 5 are government-funded private centers. A number of privately run, privately funded centers also exist. Three types of inservice training are on offer:

- Certificate training, for upgrading the teacher certificate, careers counseling certification, principalship certification, and "master teacher" certification);
- Job capacity building (e.g., subject-specific teaching methods, educational technology training); and
- Special training for teachers with a proven leadership record, which may involve enrollment in a higher degree program or international experience.

Teaching certificates are at two levels: initial teaching certificate (Level 2) and the advanced certificate (Level 1). Teachers can upgrade their initial certificate to Level 1 only after three years of teaching (or one year of teaching with a Master's degree). A Level 1 certificate is required for promotion to an administrative position. In order to obtain Level 1 certification, teachers need to take 90 hours of government-approved training, 30 hours of which comprise general courses and 60 hours in their major teaching areas. The certificate upgrade program is typically provided for three weeks during schools' summer break. Because teachers' grades in this program are entered into their promotion profile, teachers tend to work hard during it.

School-based inservice training includes teacher lesson study, consulting visits by LEA officers, and school-wide professional development. Schools are required to report school-based inservice training activities in their assessment profile, a practice that ensures all schools have some form of school-based professional development program in place. Teacher lesson studies have been the main professional development activity in South Korea for more than 30 years and have become a staple in the discourse about teacher professional development. During a lesson study, teachers observe their colleagues conducting lessons and provide feedback. A similar practice has been implemented in Japan (Lewis, Perry, and Murata 2006).

Science Teacher Professional Associations

Science teachers in South Korea have been developing their own associations in which teachers share science teaching methods, lesson designs, ideas about fun science experiments, and so on. One of the oldest such associations is the National Science Teachers Society, established in 1986 and operating under the umbrella of a progressive teachers' union (http://science.eduhope.net). The society has branches across the nation and is involved in research, course material development, professional development, running science festivals and science camps for students, and the orientation of beginning science teachers.

Another well-known association is Teachers for Exciting Science (TES, http://tes.or.kr/). Established in 1991, the association has a membership of science teachers in Seoul Metropolitan and neighboring Gyeonggi Province (about half the South Korean population resides in these two areas). Over the years, TES has transformed its purpose from developing science experiments for school science to popularizing science. Activities include a yearly science experience festival for students from kindergarten to Grade 12, science camps for science clubs across the nation, and science teacher professional development programs.

In 2003, ten science teacher groups established the Korea Science Teacher Association (KTSA), an organization dedicated to networking local science teacher groups (http://k-sta.or.kr/). TES is also a member of KSTA. KSTA initially relied on the government for funding but is also now raising funds from private companies. In 2014, a total of 18 local science teacher groups participated in KSTA activities. In addition to its networking endeavors, KSTA develops and provides science teacher professional development programs and science curricular materials for schools at all levels. It shares science lessons, including experiment-related resources, with high school science teachers and provides activities for the public and extracurricular activities for students in economically disadvantaged areas within the country and abroad (e.g., Timor-Leste).

No formal research has been conducted on the effect of science teachers' associations on student learning. However, it is obvious that teachers' associations are committed to improving school science activities and providing students with science learning experiences outside school, which would be little offered otherwise. Thus teachers serve not only within school buildings but also outside schools. What is still in question is whether teachers' participation in teacher associations has an impact on the practices and culture of science teaching in schools with respect to both students and teachers outside as well as within their membership.

Issues in Upper Secondary Science Teacher Quality

Teacher Supply

Traditionally, respect for teachers in Korea has been rooted in Confucian philosophy in which monarch, teacher, and parents are accorded respect. This cultural value decreased with industrialization and capitalism. However, teaching as a job regained its popularity as a result of the economic crisis of 1997, which greatly reduced job prospects and meant job stability could no longer be taken for granted except in the public service sector. Teaching in schools, whether public or private, is a secure job in that teachers are tenured upon hiring until they reach legal retirement age (62 years). The popularity of teaching guarantees an adequate supply of quality teachers. The OECD TALIS report (2014) highlighted this situation when it recorded that 90 percent of secondary school teachers in South Korea teach in their certified fields compared with the OECD international average of 73 percent.

According to a survey conducted by the Korea Research Institute for Vocational Education and Training in 2012, elementary school teaching is the occupation that secondary school girls most want to enter, and secondary school teaching is the third. Among secondary school boys, elementary school teaching is the seventh most popular occupation and secondary school teaching the eighth. Accordingly, high-achieving students apply for admission to teacher education programs. The popularity of teaching as a profession might also reflect teaching's employment conditions compared to other occupations (Kim and Han 2002). The OECD report (2014) positioned South Korean secondary teachers' average salary as 36 percent higher than that of full-time workers with a degree as of year 2012. While the beginning salary for teachers in South Korea is lower than the OECD average, after 15 years of teaching a teacher can expect to be earning a salary that has increased by an average of 76 percent, which makes it the fifth highest among the 36 countries surveyed.

Those individuals who have attained teaching certificates must take national teacher employment tests to obtain employment in a public-sector school. The public school teacher employment test has evolved over the past 30 years. The most recent form of the test consists of two parts: a pen-and-paper test and a lesson demonstration with interviews. This system applies to both elementary and secondary teacher employment. Twenty percent of the pen-and-paper test for secondary science teaching applicants consists of educational foundation questions (e.g., classroom management, educational psychology). The remaining 80 percent focuses on subject-matter content and pedagogy. Until recently, test questions were multiple choice in format but now comprise short-answer and essay questions.

The paper test is administered nationally, but the lesson demonstration with interviews part is conducted by city and provincial education offices. Applicants are assessed by an office in the area of the country where they want to teach. During the lesson demonstration component, applicants are given a lesson topic and lesson activities and asked to use that material to develop a plan for a one-hour lesson. They are then required to demonstrate specific parts of the lesson in front of several examiners for 20 minutes. Some cities and local education offices also require applicants to perform a one-hour experiment as if they were a student. On completion of their demonstration work, applicants are interviewed about their educational philosophy and other general issues. Applicants can expect to be asked questions relating to their character and disposition toward teaching during the interviews.

The popularity of teaching is also shown in the teacher employment test. The ratio of total to successful applicants for teaching mathematics, Korean, and English is typically around 30 to 1 while that for teaching sciences (physical, chemistry, biology, Earth science) is typically 9 or 10 to 1. However, these competition ratios vary across cities and provinces. The number of positions available in each city or province is announced a couple of months before the test, and applicants apply based on the vacancies and the areas of the country where they would prefer to teach. Teachers who pass the test are assigned to schools within a month or two.

Although private schools use their own procedures to hire teachers, South Korean law requires the qualifications and status of these teachers to be comparable to those of teachers working in public schools. Private schools typically announce vacancies through teacher education programs that also serve as a means of recruiting teachers, and they rarely use teacher employment tests in their hiring processes. In general, public school positions are more popular than those of private schools because teachers in the former are deemed to have a freer rein.

Teacher retention has not been an issue in South Korea, as there is a very low turnover rate (Korean Education Statistics Service 2015). In 2014, for example, there were 55,000 teachers in both public and private general high schools. Of these teachers, 853 (1.5 percent) retired and 243 (0.4 percent) resigned during the year while 1,618 (3.0 percent) were newly hired teachers. Teachers rarely leave their job. However, while teachers in private schools tend to stay in the same school, public school teachers rotate their schools every four or five years (for details, see Kang and Hong 2008). The rotation system ensures that all students have equal access to various teachers and all teachers have equal access to various schools.

Teachers' promotion profile ratings are often compared when applications for teaching positions exceed vacancies. Newly hired teachers do not have the right to choose which level of school they work at. However, teachers can apply to change school levels at their regular rotation time. New secondary teachers may be assigned to a middle or high school depending on vacancies. Those teachers who begin their professional lives in a middle school usually keep rotating between middle schools, and those who first work in a high school normally keep rotating between high schools. Teachers are asked to indicate three schools of their preference in rank order at the time of their rotation but may still be assigned to a school not on their list. Teachers working in schools designated as "difficult to work at" need to stay in that school for only three years, and during this time additional points are added to their promotion profiles. Thus, schools that are "difficult" are popular among teachers who want high promotion profile ratings.

Professional Education and Training Issues

The fact that preservice teachers are expected to undertake the employment test brings relative uniformity and standardization to South Korea's various teacher education programs. The employment tests are developed by expert teachers as well as individuals with expertise in pertinent fields, such as educational foundations, sciences, and subject-specific content (e.g., physics education). Together, teacher education academic personnel, teachers, and disciplinary experts shape the test and thereby influence the curriculum of teacher education programs. Those constructing the tests are aware that subject-matter content questions can only be pitched at a certain level so that the examinees can answer multiple questions in the given time. The topics featured in these tests recur, so students tend not to take elective courses that do not address those topics. Some teacher education programs tailor their contents to the test, a situation that lessens teacher preparation diversity and high content standards. This is even more obvious in the lesson demonstration part of the test because most lessons demonstrated by applicants are similar. Applicants therefore practice their lesson demonstrations in accordance with the content of previous test tasks and assessment frameworks.

Very few cities or provinces provide initial induction programs; those that are provided are typically only a day long (see, for example, Seoul Metropolitan Office of Education 2014) and focus on the duties that come with the job rather than on the actual teaching. In general, beginning teachers are expected to start out in the classroom by applying what they learned during their student teaching and then to learn on the job from there.

The fact that secondary science teaching certificates are subject specific creates a problematic gap between teacher preparation and the science curriculum at the middle school level. The middle school science curriculum offers science in the form of general science, which combines topics from all four science fields (physics, chemistry, biology, and Earth science), but of course secondary science teachers are trained to teach a specific science discipline offered at the high school level. The same problem is likely to occur when South Korea implements its new high school curriculum because it requires all students to take the one integrated science subject.

Issues at the Chalkface

Choices within the high school curriculum are heavily influenced by CSAT, the national college entrance examination. Students select courses that they will be examined in; basic science courses are more popular than advanced courses because of the higher scores they bring. Likewise, those science fields seen as easier to get higher scores in are the more popular. For the CSAT administered in 2014, 61 percent and 58 percent of applicants chose biology I and chemistry I respectively, whereas only 23 percent of applicants selected physics I. In order to meet graduation requirements, science-track students also take elective science courses that are not part of their CSAT examinations. Students are less motivated to learn these subjects, which presents teachers with the challenge of engaging students in that learning. In the worst-case scenario, which according to teachers does happen sometimes, teachers cover their subjects briefly and then let students study those of their subjects that will be CSAT tested.

Hong, Song, and Kim (2009) compared how South Korean, Japanese, and US teachers were teaching elective science courses at Grade 11 and 12 levels. During their study, the three researchers observed classroom science lessons, surveyed students about the classroom environment in terms of their level of engagement in those lessons (see also in this regard Weiss et al. 2003), and interviewed teachers about their teaching practices in relation to those listed in the student survey. As would be expected in a system where a high-stakes test shapes curriculum implementation, the study revealed that teachers typically lectured to cover the material, rarely used thought-provoking questions and student-based activities, and rarely offered laboratory activities. Overall, the researchers saw little evidence of meaningful interactions among students or between teacher and students. Given this type of science classroom environment, the researchers were not surprised to find a low level of engagement among the students.

South Korea's government requires teachers to assign 30 percent of a student's science course grade to his or her performance on practical (e.g., laboratory) activities. However, some teachers use pencil-and-paper rather than practical activities for performance assessment. Because school examinations totally rely on teacher-developed tests, teachers can shape their courses and tests as they wish. Teachers' reluctance to use "real" performance assessment is partly because of their lack of confidence in ensuring the reliability and validity of their assessments. Also, the need to cover all the materials likely to be covered by the CSAT test and to prepare students for it deprives teachers of the liberty of taking class time to try out meaningful learning activities.

Continuing Professional Development Issues

Over the past ten years, South Korea has seen an increase in calls for targeted teacher inservice training because of the public becoming increasingly critical of what it sees as the lack of responsiveness of schools and teachers. A phenomenon called "school collapse" not only represents the growing general distrust of public school education (Kim 2004) but also has persuaded the government to create inservice training requirements for continuing teacher development. School collapse is characterized by student disengagement in lessons and teachers unable to manage student behavior effectively.

A common discourse related to the school collapse phenomenon is that students in the digital era require classroom management and teaching methods that differ from those of the past. Implementation of these new methods is seen as even more important in a country such as South Korea where high-speed Internet access reached 100 percent of the population in 2011, an achievement that gave the country the number-one ranking on this index among OECD countries. In 2013, 96 percent of secondary school students reported that they used the Internet for leisure (Korean Educational Statistics Service 2014). The need for teachers to upgrade their management skills and teaching methods is therefore obvious. Since 2008, the government has required teachers to participate in 90 hours of professional development courses every three years after four years of teaching. Of the 90 hours, 60 hours have to focus on subject-specific areas or "student-consulting" activities, such as involvement in school violence programs, career guidance, and extracurricular events. This complement of professional development is additional to school-based professional development.

A prominent area of inservice training for science teachers concerns science experiments (e.g., via a microcomputer-based laboratory or through other lab-based digital gadgets) and teaching methods. LEAs regularly offer science experiment training for 30 hours during school breaks in summer and winter. While the newly instituted requirement for teacher inservice training is an improvement, this quantitative focus on training now needs to be superseded by quality assurance (Park 2014). At present, the government's teacher professional development policy lacks a means of assessing the effectiveness of teacher training programs, so it is understandable why the call for quality inservice training has become increasingly louder.

Trends and Developments in Upper Secondary Science Teacher Quality

Upper secondary science teacher-quality issues can be summarized in terms of the current inservice training system, the education system in general, and the educational culture of South Korea. Quality teacher supply is not a problem in South Korea, so what is critical is the need to keep developing teacher quality during teachers' tenure. Until now, inservice training has increased in quantity. However, as mentioned above, it is now time to review its quality.

South Korea's inservice training system has been a mix of bottom-up and top-down approaches. Inservice training such as school-based teacher-initiated training and teacher professional association-initiated training is bottom-up; professional development courses offered by central or local LEAs to fulfill government requirements are top-down. There is currently insufficient external evaluation to identify how much these types of inservice trainings influence teachers' job satisfaction and classroom teaching practices. Given the increasing call for quality assurance, rigorous external evaluation processes need to be instituted in order to provide teachers with more evidence-based professional development programs.

Even with an adequate quality teacher supply, beginning teachers still need initial teacher induction programs so they can further their practical knowledge and gain professional expertise (Feiman-Nemser 2001; Schneider and Plasman 2011). Recent government professional development plans emphasize the need for new teacher induction programs. In 2013, about 60 percent of all beginning teachers offered participation in an induction program took up that offer. This rate of participation was about 10 percent higher than the OECD average (OECD 2014). However, the quality of these induction programs received little attention. Furthermore, in South Korea, the programs are provided some time after teachers have started teaching in the classroom, leaving first-year teaching as a "blind spot." A more thoughtful investment in beginning teacher induction programs needs to be considered.

The teacher preparation system is poorly aligned with the 6-3-3 education system and its respective curricula. Secondary science teachers are prepared for specific subjects such as physics and biology. However, as mentioned earlier in this chapter, the majority of these teachers teach in middle schools where science is an integrated subject. This gap has been recognized since the 1990s when, in an attempt to address the problem, some secondary teacher education programs created a general science education minor. However, because the teacher employment test does not offer a general science option, few aspiring teachers take the minor. The current curriculum reform movement in which eight credits of integrated science will be required for high school graduation and for the CSAT will only widen this gap in teacher preparation. While middle school general science presumably causes little difficulty for teachers prepared in only one science field, high school integrated science will exacerbate the problem of teachers teaching out of field. How this issue is resolved remains to be seen when the new curriculum is implemented after 2017. Perhaps inservice training for content knowledge will be provided while teachers in schools might turn to team teaching where they can.

South Korea's high-stakes test-oriented educational culture causes a particularly difficult moral dilemma for teachers. Students and parents want teachers to teach to the CSAT, and students enrolled in non-CSAT courses are disinclined to engage in them. In endeavoring to resolve this issue, the government keeps encouraging colleges and universities to increase the weighting of academic grades in relation to CSAT scores in their admission screening. However, the fact that schools have student populations of different quality makes grade point average a less valid criterion than a common test. Furthermore, using CSAT scores is a cost-effective and thus expedient means of screening students. There is a tug-of-war between the government and tertiary education institutions: the government seeks to make CSAT easy enough so that it measures only minimum scholastic aptitude, while universities and colleges seek to make full use of CSAT scores when selecting students and therefore want the test to differentiate between them. In this context, typical high school science teachers feel obliged to ensure that their university-bound students can enter institutions of their choice, which forces these teachers to teach to the test rather than offer meaningful science learning experiences. This focus on tertiary education entrance, which pervades South Korea's entire educational culture, needs to be transformed.

References

Feiman-Nemser, Sharon. 2001. "From Preparation to Practice: Designing a Continuum to Strengthen and Sustain Teaching." *Teachers College Record* 103 (6): 1013–55.

Hong, M., H. Song, and J. Kim. 2009. *A Study on Science Classroom Learning at High School in Three Countries: Korea, United States and Japan.* Seoul: Korea Institute of Curriculum and Evaluation.

Kang, Nam-Hwa, and Miyoung Hong. 2008. "Achieving Excellence in Teacher Workforce and Equity in Learning Opportunities in South Korea." *Educational Researcher* 37 (3): 200–07. doi: 10.3102/0013189X08319571

Kim, Jeong Wan. 2004. "Education Reform Policies and Classroom Teaching in South Korea." *International Studies in Sociology of Education* 14 (2): 125–45.

Korean Education Statistics Service (KESS). 2014. *Statistics on Adolescents.* Seoul: Author.

———. 2015. *Annual Turnover of Teachers.* Seoul: Author, http://kess.kedi.re.kr/index.

Kim, Ee-gyeong, and You-Kung Han. 2002. *Attracting, Developing and Retaining Effective Teachers: Background Report for Korea.* Seoul: Korean Educational Development Institute.

Lewis, Catherine, Rebecca Perry, and Aki Murata. 2006. "How Should Research Contribute to Instructional Improvement? The Case of Lesson Study." *Educational Researcher* 35 (3): 3–14. doi:10.3102/0013189X035003003

Organisation for Economic Co-operation and Development (OECD). 2014. *TALIS 2013 Results: An International Perspective on Teaching and Learning.* France: OECD Publishing.

Park, K. Y. 2014. *The Current Status and Future Directions of Inservice Training for Teacher Professional Development: An Examination of Current Issues,* http://edpolicy.kedi.re.kr/EpnicForum/Epnic/EpnicForum01Viw.php?Ac_Code=D0010102&Ac_Num0=17180 (in Korean).

Schneider, Rebecca M., and Kellie Plasman. 2011. "Science Teacher Learning Progressions: A Review of Science Teachers' Pedagogical Content Knowledge Development." *Review of Educational Research* 81 (4): 530–65. doi: 10.3102/0034654311423382

Seoul Metropolitan Office of Education. 2014. *New Teacher Training and Student Guidance.* Seoul: Author, http://m.newspago.com/a.html?uid=28615 (in Korean).

"Test-Taking in South Korea: Point Me at the SKY." *The Economist,* November 8, 2013, http://www.economist.com/node/21589573.

Weiss, Iris R., Joan D. Pasley, P. Sean Smith, Eric R. Banilower, and Daniel J. Heck. 2003. *Looking Inside the Classroom: A Study of K–12 Mathematics and Science Education in the United States.* Chapel Hill, NC: Horizon Research Inc.

Further Reading

Burt, Matthew E., and Park Namgi. 2009. "Education Inequality in the Republic of Korea: Measurement and Causes." In *Inequality in Education: Comparative and International Perspectives,* edited by Donald B. Holsinger and W. James Jacob, 261–89. Netherlands: Springer.

Jones, Randall S. 2013. *Education Reform in Korea* (OECD Economics Department Working Papers, No. 1067). Paris: OECD Publishing, http://www.oecd-ilibrary .org/economics/education-reform-in-korea_5k43nxs1t9vh-en.

Kim, Chung Wah, and Myron H. Dembo. 2000. "Social-Cognitive Factors Influencing Success on College Entrance Exams in South Korea." *Social Psychology of Education* 4 (2): 95–115.

Kim, Meesok. 2003. "Teaching and Learning in Korean Classrooms: The Crisis and the New Approach." *Asia Pacific Education Review* 4 (2): 140–50.

Paik, Susan J. 2001. "Introduction, Background, and International Perspectives: Korean History, Culture, and Education." *International Journal of Educational Research* 35 (6): 535–607.

———. 2004. "Korean and US Families, Schools, and Learning." *International Journal of Educational Research* 41: 71–90.

Park, Hyunjoon, Soo-yong Byun, and Kyung-keun Kim. 2011. "Parental Involvement and Students' Cognitive Outcomes in Korea: Focusing on Private Tutoring." *Sociology of Education* 84 (1): 3–22.

Seth, Michael J. 2002. *Education Fever: Society, Politics, and the Pursuit of Schooling in South Korea.* Honolulu: University of Hawaii Press.

Sorensen, Clark W. 1994. "Success and Education in South Korea." *Comparative Education Review* 38 (1): 10–35, http://storage.globalcitizen.net/data/topic/knowledge /uploads/20131115119479333885_education.pdf.

9

Cluster Summary One

The preceding seven countries are all high-income OECD members with long-established, stable education sectors (reforms notwithstanding). The status of upper secondary schooling varies considerably from being part of a largely unified post-primary level of education as in Australia to being a discrete institutional entity as in Singapore (where it is called "pre-university"). Science education in these fortunate societies is well developed and is well resourced with well-defined science specialization tracks available to upper secondary students, particularly in the Asian and European members of this set. With regard to upper secondary science teacher quality, the following transpire as common themes.

Teacher Supply

Teaching in this cluster is generally an attractive profession from the point of view of remuneration, social status, and working conditions. In the Asian countries, in particular, places in post-primary teacher preservice programs are sought-after pre-employment destinations. Where the upper secondary level forms a discrete layer of the education pyramid, the standing of teachers at this level is commensurate. However, the profession has to compete with other career pathways, and this particularly affects the supply of teachers in the physical sciences, graduates of which have viable alternatives they can avail themselves of. Ageing secondary science teaching forces appear to be a feature of most of these systems, and the profession may have to be made more enticing in the foreseeable future to young graduates. At the same time, the demographic transitions in these countries are seeing a decline in science teacher demand.

Out-of-field teaching is a common phenomenon in science education at the lower secondary level where teachers often have to deliver units in fields outside their own specialization (e.g., an Australian biology graduate teaching chemistry, physics, and Earth science topics as part of a junior-level "general science" course), but it becomes a problem at the upper secondary level where a graduate with a specific science-subject expertise ideally needs to be in front of every class. One way of addressing chronic shortages of physics teachers is to recruit graduates and mid-career professionals in associated fields such as engineering. There are also the perennial issues of staffing schools in remote and "undesirable" areas, which systems address in their own ways. These issues are not exclusive to upper secondary science but may be all the more acute in the context of shortages of specialists in physical sciences.

Teacher Academic and Professional Education and Training

On the whole, upper secondary science teachers in these countries are well qualified and well trained. Whether the track a typical practitioner followed is of the "'concurrent" or "consecutive" type (i.e., subject qualifications and teacher training as part of a single degree program or teacher training on the heels of a subject degree), he or she will have degree-level qualifications in science and undergone pedagogical training that includes field practicums. Where upper secondary schooling constitutes a distinct educational tier, subject degree requirements may rise above the Bachelor's level; advanced certification involving the acquisition of additional qualifications may also be required.

Continuing professional development (CPD) is a ubiquitous feature of teaching in these countries, with teachers being entitled to (in some cases, obliged to) undergo CPD on a regular basis. The nature of inservice training provision varies considerably, and its effectiveness is infrequently questioned. It appears to be particularly well developed in the Asian countries, where it is highly focused and purposive and there is an expectation that teachers will put their CPD-refined skills into practice in the classroom. The discernible move in countries with well-developed CPD systems toward a more seamless transition from "training" to "practice" is resulting in the two operating more in tandem throughout a practitioner's career.

Issues at the Chalkface

Resource constraints are a constant source of dissatisfaction for practitioners of all professions, and teaching is no exception. It may nonetheless be said that the availability of laboratories and the requisite equipment for the teaching of upper secondary science is not particularly problematical for science teachers at the upper secondary level. The availability of technical assistance is not quite on a par, though, with teachers in some countries often finding themselves without. This lack of assistance can lead to the neglect of practical work. However, a recurring theme is that of practical work playing second fiddle to theoretical teaching, especially in the context of high-stakes, examination-driven curricula. But even where there is plenty of practical work occurring, it tends to be of the "cookbook" style rather than the genuinely experimental or inquiry based.

It should be recalled, however, that the principal function of upper secondary schooling from students' and parents' points of view is that of a bridging mechanism from school to tertiary education, and it is examination performance that largely determines the outcome of schooling at this level. Given the content-intensive nature of most upper secondary specialized science curricula, it is hardly surprising that teachers at this level are inclined toward more conventional teaching methods (direct instruction and verification-type laboratory work) instead of the dictates of constructivist models and inquiry-based approaches. Yet this pattern still tends to dominate, even where the impact of external examinations has been lessened. At the core of this issue lies a deep-seated ideological disagreement that is unlikely to be reconciled. The Dutch approach—offering a choice between conventional "cognitivist" and Brave New World "constructivist" science education—represents a pragmatic response to this irresolvable impasse.

Science Teacher Professional Associations

A striking feature of these varied education systems is the profile of professional associations that science teachers can join. Science teacher associations, especially at the national level in these countries, not only represent their members' interests but have become prominent actors in science education academia and/or teacher professional development. It is these associations that have put paid more than anything else to Shaw's criticism of teachers being "those who can't" and that have elevated teachers, particularly highly qualified upper secondary teachers, to the same professional heights as practitioners of other professions.

10

Oman

Abdullah K. Ambusaidi

Teaching Science in the Sultanate of Oman

Backdrop

The Sultanate of Oman is located in the extreme southeastern corner of the Arabian Peninsula. It has a total land area of about 300,000 square kilometers and a population approaching four million, of whom about 56 percent are Omani citizens; expatriates account for about 44 percent of the population. The Sultanate of Oman occupies a privileged geographical position between the countries of the world, making it a focus of trade between East and West. Omani society, like many other societies in the region, is witnessing profound changes across all spheres of life, including education.

The Omani government sees education as a vehicle for development and for leading society toward modernity. Schools have been built all over Oman, regardless of existing variation among the different governorates of the Sultanate. These schools have all the requirements, including qualified teachers, curricula, and teaching, needed to provide students with quality education.

The Omani school education system consists of 12 grade levels, divided into basic education (Grades 1 to 10) and post-basic education (Grades 11 and 12). Basic education is divided into two cycles—one from Grades 1 to 4 and the second from Grades 5 to 10. Kindergarten is not part of the formal education system. Students usually enter school at age six.

The Ministry of Education does not have official statistics on the transition rate from basic education to post-basic education. However, according to Al-Rasbi et al. (2012), the proportion of students in 2012 who failed one or more subjects in Grade 10 was about 10 percent. Accordingly, we can assume that, on average, the percentage of students not transitioning from basic education to post-basic education is about 10 percent of the cohort.

Post-basic education is a very important stage for students because it leads to higher education or the world of work. The Ministry of Education has conducted several consultative studies on post-basic education (Grades 11 and 12) with the aim of identifying and developing the skills that students in these grades need to attain levels of academic performance that meet international standards. The ministry has also conducted a number of seminars to reinforce post-basic education, such as one organized in cooperation with UNESCO and titled Education and 21st Century Competencies. Held in September 2013, the seminar's theme focused on the post-basic skills required for the twenty-first-century labor market (Ministry of Education 2014).

School Science

Science is one of the basic subjects in the Omani education system. It is taught from Grade 1 to Grade 12. Students in Grades 1 to 10 are allocated one science textbook per year level. The textbooks include content on the three main fields of science (biology, chemistry, and physics) along with some Earth and space science. In upper secondary schools (Grades 11 and 12), students work from separate textbooks for biology, chemistry, and physics. Students can choose to study either all three subjects or two of them if they intend to major in science-oriented specializations at the tertiary level of education. They can study only one science subject, usually "science and technology" in Grade 11 and "science and environment" in Grade 12, if they want to pursue arts-oriented specializations at the tertiary level.

The weekly number of lessons allocated to each science subject in upper secondary schools is four. The major topics covered in Grade 11 biology are living matter, diversity and adaptations, internal transport, and plant biological processes. The major topics in Grade 11 chemistry are periodicity, acids and bases, stoichiometry, and organic compounds. Grade 11 physics covers motion and dynamics, conservation of energy and momentum, and electricity. In Grade 12 biology, the areas of emphasis are cells, the nervous and endocrine systems, and human reproduction and heredity; in chemistry, they are electrochemistry, reaction, the gas laws, and chemical equilibrium; and in physics, electromagnetism, acoustics, electromagnetic waves, and atomic physics.

There is a final examination for all science subjects in both semesters for Grades 11 and 12. The Directorate General of Educational Evaluation at the Ministry of Education's head office prepares and constructs the examinations according to ministry specifications. The language of the examinations for public schools is Arabic; in private schools (bilingual schools), the language is English. The content of the examinations in bilingual schools differs from the content of the public school examinations due to the differences between the curricula in the two school types. International schools follow external curricula and examination prescriptions such as those of the Cambridge International Examinations. Students at public schools receive the General Diploma in Education if they successfully complete Grade 12. Their post-school options depend on their diploma marks. Higher-performing students have opportunities to secure places at good universities and colleges inside and outside Oman. Students with low marks often find useful employment in various public- and private-sector occupations, entry to which is not reliant on high academic standards. For example, many students with low marks enter the army or police force or work for oil companies.

Assessment of student learning of science has undergone major developments and now incorporates marks obtained on school-based assessments. These account for 40 percent of the total achievement score in Grade 11 and 30 percent in Grade 12 (Ambusaidi and Al-Balushi 2014). Science teachers are required to use several tools to assess students, including oral examination, student presentations, homework, practical work (scientific activities and practical examinations), and classroom tests. The centrally designed final examination and the tests constructed by schools must contain questions that cover three cognitive skills: retrieving (mainly recall based), applying (e.g., using models, finding solutions, interpreting data), and combining (higher-order cognitive).

A practical examination is implemented at the end of each semester of the academic year for both Grade 11 and Grade 12 for the three science subjects. Students take this exam individually. Each student is given an experiment and asked to perform it under a teacher's observation. The experiments must not be the same as those that students carried out during the semester. Instead, they need to be new ones in terms of materials but to cover the concepts underpinning the experiments conducted during the semester. Teachers are required to use an observation sheet based on marking keys and itemizing the skills that students should be able to demonstrate, such as their abilities to plan and conduct experiments and to interpret and analyze the results.

Moderation is used to achieve and ensure consistency across teachers, both in the use of criteria and in the application of standards. Inspectors and supervisors are an integral part of this process. Each regional office of the Directorate General of Education has a committee formed of supervisors of school subjects visit each school with Grades 11 and 12 in their region. The committee's task is to ensure that teachers have implemented the assessment tools and standards according to ministry specifications. The committees also make sure that each student's given marks reflect his or her real performance and have not been inflated by teachers.

Teacher Academic and Professional Education and Training

Currently, Sultan Qaboos University is the only government university responsible for preparing science teachers. The College of Education at the university offers teacher education programs for all subjects, including science. The main aim of the college's science program is to prepare Omani science teachers to teach the three sciences (biology, chemistry, and physics) at Grades 5 to 12 levels. The subject-matter courses on offer through the college's science education unit are delivered in English. The College of Education itself is responsible for the educational and pedagogical courses that students need to take, and these are delivered in Arabic. Students also take science teaching methods courses in their third and fourth years of study. Each course involves two contact hours a week for theory and another two contact hours during which students put into practice what they have learned in their theory classes. The practice component takes place in peer teaching settings (i.e., microteaching) and on site (in schools). Students undertake teaching practicum, also called field experience, in their final year once they have completed 90 percent of the courses making up the degree. These prospective teachers also attend school five days a week in order to practice all school-related work, including administrative tasks as well as actual teaching.

Continuing Professional Development

The Omani Ministry of Education is committed to playing a key role in continuing professional development (CPD) for teachers so as to enhance and develop the quality of teaching and assessment processes in the Sultanate. The Directorate General of Human Resources at the ministry is responsible for planning, organizing, and conducting professional development programs both centrally and across the 11 directorates of education in Oman. In 2013, His Majesty Sultan Qaboos bin Said increased the annual budget for professional development programs by seven million Omani rials (around US$18 million). The CPD programs on offer vary in terms of subject-matter content and pedagogy content. Programs include the following:

- *Enrichment programs for experienced teachers:* This program is a collaboration between the Ministry of Sultan Qaboos University. It targets teachers with four or more years of teaching experience and is implemented twice a year, in January and June, over two weeks. Its main focus is subject matter, but 10 percent of the total program hours are given over to pedagogy. The number of science teachers participating in each session is, on average, around 240 from the various educational regions.
- *Training program for new teachers:* Oman recognizes that CPD continues on from initial training and that mechanisms must be in place for supporting individuals throughout their initial teacher training and beyond (Bell 2006). The Omani Ministry of Education accordingly offers a program that is not only aimed at preparing teachers academically and pedagogically to teach in schools but also at keeping new teachers updated with developments occurring in the Omani education system. The program is implemented centrally by the ministry's head office or by the regional Directorate General of Education offices (Ministry of Education 2013a). Teachers also receive information about administrative matters affecting schools.

- *Incentive program for teachers:* This program is conducted annually. Its aim is to develop teachers' knowledge and skills in ways that enhance their professional efficacy in such areas as self-development, working with groups, and communication in the working environment. The ministry coordinates this program, which is conducted by specialists from the Arab world and in all 11 directorates of education. To date, the ministry has evaluated two rounds of the program in order to determine and then make needed improvements to it.

- *Teachers and supervisors conference participation:* CPD goes beyond attendance at workshops or training programs. It also includes participation in local and international conferences. An example of such a conference was the Seventh Conference on Science, Mathematics, and Technology Education, organized jointly by Sultan Qaboos University and Curtin University of Technology (Australia) and held in November 2013. The Omani Ministry of Education sent more than 300 science, mathematics, and technology teachers and supervisors to the conference.

- *Continuous short-term programs:* Teachers attend these programs either before the beginning of the school year or during it. Each program usually lasts a maximum of one week and tackles issues directly related to teaching or assessment. The programs are of two types: those planned and conducted centrally, and those planned and conducted regionally (by the education authorities of the regions). The first type of program targets representatives of teachers or supervisors from throughout Oman. Participants are expected to share what they have learned with their counterparts in the regions. The second type usually targets a larger number of teachers in each region who can then transfer what they have learned in the program to their colleagues at school. Examples of training programs for upper secondary science teachers include those targeting specific science subjects (e.g., biology experiments), assessment-related skills (e.g., item construction), and general classroom management (Ministry of Education 2013b).

- *Peer visits within and between schools:* In order for teachers, particularly new ones, to benefit from their teaching colleagues' experience, especially that of expert teachers, the ministry encourages teachers to visit one another in their classrooms. The principal of each school arranges the visits. After each visit, teachers discuss what they have learned and observed.

- *The school as a learning community:* As the result of a ministry initiative, each subject in each school is overseen by a so-called "first teacher" or "principal teacher." These teachers are required to perform roles directed toward improving the quality of teaching in their schools. These roles include giving support to colleagues, visiting teachers for formative and summative purposes, planning and arranging workshops, holding regular meetings with other teachers to solve problems and develop teaching of the subject, and arranging peer visits within and outside schools. The ministry relies heavily on these principal teachers to realize its aim of higher-quality teaching in schools.

Teachers in Oman can also access opportunities for further professional development by conducting research, especially action research. Teachers conducting the best such research receive prizes. The focus of the research is mainly on teachers reading in the area of interest and developing instruments that might help improve the quality of their teaching. The Ministry of Education is currently running a program in collaboration with Sultan Qaboos University to train a number of teachers on how to use action research to solve school-related problems.

Science Teacher Professional Associations

Although there are no professional educational associations for Omani science teachers, there is one official association that science teachers can join and benefit from—the Omani Society for Educational Technology (OSET). OSET is a non-profit national organization that was established in 2007 by ministerial decree 39/2007. The society's vision is to advance educational technology in Oman by developing strategies for the planning, implementation, and successful application of educational technology. To help them achieve this aim, the society draws on the technical skills and resources of governmental and private educational organizations and individuals with expertise in the field of educational technology. OSET activities include exchanging experiences, issuing newsletters, setting up training workshops, engaging in research, building an electronic database, and conducting public lectures and conferences. Even though science teachers can benefit from these activities, there is an urgent need for them to have their own dedicated association of the kind found in other countries, such as the National Science Teachers Association in the United States and the Association of Science Education in the UK.

Issues in Upper Secondary Science Teacher Quality

Teacher Supply

Teachers in Oman normally progress to upper secondary schools after initial experience and upgrading of qualifications at the lower level of schooling. However, in some cases, new graduates immediately begin their teaching career in post-basic education (upper secondary schools). This situation is particularly common among biology teachers and those teachers assigned to work in schools with a low number of students, generally located in mountainous and desert regions of the country. However, the problem facing the ministry is that many good teachers in basic education schools refuse to teach at post-basic education schools not only because of the extra workload and pressure that teaching at that level entails but also because there is no incentive for them to take on this greater degree of responsibility. Students and parents consider the upper secondary grades a crucial part of the education system because graduation from them leads to higher education. Science and mathematics are two very important subject areas at this level, and teachers have to work hard to ensure that their students will get good marks in the General Diploma Certificate.

The Ministry of Education works to recruit Omani citizens to work as teachers in both basic and post-basic education. These teachers are drawn from public and private universities inside and outside Oman. The ministry has no major concerns about the quality of teachers who graduate from universities inside Oman, but they do have concerns about the teachers who graduated outside the country. Some of these latter teachers lack in-depth subject knowledge, and some do not have enough training as science teachers in schools. The ministry's efforts to solve this problem include development and implementation of the inservice training programs described earlier in this chapter.

Another problem additional to teacher quality is teacher shortage, especially of male teachers of science subjects but mainly biology. One reason for this is the freezing of the science education program at Sultan Qaboos University for two years (2010 and 2012) and the closing of the program at other governmental and private institutions following inaccurate estimates from the Ministry of Education and the Ministry of Higher Education about the number of students likely to study science education. Another reason is that the salaries of teachers compare poorly with the salaries for other professions, such as medicine or engineering, and a third reason is the low social image of teachers. A fourth reason relates to the many scholarships the government offered Grade 12 diploma graduate students during the three years encompassing 2012 to 2014. These scholarships enabled their recipients to study outside Oman in specializations other than education. Many male students applied for these scholarships, which resulted in fewer male students entering teacher education institutes.

One solution to the problem of teacher shortage that the ministry has adopted is recruiting teachers from other Arab countries. However, while this initiative may solve part of the problem, it has had negative effects on student achievement. Although these teachers are Arabic speakers, students have difficulty understanding some of them, especially those from Tunisia. Also, most of these teachers are placed in schools far from the major cities, and they find adapting to working in rural areas difficult. Services such as health care, large markets, and entertainment venues are usually limited in these places. Some of these teachers resign after a relatively short time, a situation that naturally affects students' learning until a suitable replacement is found. Yet another issue concerns the extent to which these teachers understand the education system in Oman. Most arrive in Oman just two weeks before the school year starts, so they do not have time to gain in-depth understanding of how the education system works. Some refuse to attend training programs or to respond to supervisors' comments and suggestions on ways of improving their teaching. They may see themselves as experts in the field and therefore not needing of further development.

Professional Education and Training Issues

Most teachers teaching at Grades 11 and 12 spent at least five years teaching in basic education schools (Grades 5 to 10) before being upgraded to teach at the upper secondary level. Oman's government offers Omani science teachers opportunities to strengthen their mastery of their subject-matter knowledge, pedagogical content knowledge, or knowledge in related areas such as educational psychology or educational technology by supporting them financially to pursue a higher degree (Master's or PhD) inside or outside Oman. The funding covers both tuition fees and living costs. Each year, the Ministry of Education sends more than 200 teachers, supervisors, and administrators to Sultan Qaboos University to study for higher degrees. In order to qualify for this study, teachers and supervisors need to have at least four years of experience in teaching, to meet academic requirements such as a cumulative grade point average of 2.5, and to have passed both the entrance examination and the interview. One issue raised in regard to the Master's program in science education at Sultan Qaboos University is that most students are female. Some cohorts have had no male students, and yet almost 50 percent of science teachers in Grades 5 to 12 are male.

Issues at the Chalkface

Developments in the content and nature of the post-basic education science curriculum have been accompanied by changes in teaching and learning strategies (Ambusaidi and Al-Balushi 2015). Although science teachers have the freedom to use the teaching method or strategy they want, the Ministry of Education continues to put a great deal of emphasis on student-centered teaching and learning strategies (Ministry of Education 1998). The ministry encourages teachers to use two main teaching methods that reflect student-centered teaching and learning—inquiry-based learning and cooperative learning. However, teachers often experience difficulty when using inquiry-based learning because they cannot readily implement it for each individual student owing to the large number of students per class (35 students on average), shortage of time, and lack of materials. In an effort to address this issue, the ministry requires teachers to implement inquiry-based learning in groups of four or five students. However, this approach raises questions about whether or not each student actually has the opportunity to practice inquiry activities and not be dominated by other students in his or her group.

A second difficulty is that of being able to condense the large amount of textbook content to be covered in each semester. As noted earlier, Grades 11 and 12 are the crucial grades for students and their parents because their goal in attending upper secondary schooling is to score high marks in the internal and external examinations. This expectation forces teachers to concentrate on covering all the content in the textbook using the most direct methods of teaching, such as "in front of the class" demonstrations and semi-lecturing.

But quality education at the upper secondary level requires methods of teaching and learning that include graphic organization, reading in order to find information (research), role-play, narratives, and classroom discussion. These types of learning activity (especially graphic organization for note-taking, research, independent use of textbooks, and classroom discussion) are essential skills for students studying at tertiary level. If the ministry is to truly encourage the use of other types of teaching methods in upper secondary science, it needs to review the amount of content teachers are expected to cover or consider allocating extra lessons per week for each science subject.

One problem that has arisen in relation to the practical examination is that students in some schools do not conduct experiments by themselves during the semester, which means they do not grasp the skills they will be assessed on in the practical examination. Generally, teachers conduct the experiments, and students are just spectators with little participation. Students complain that the practical examination turns out to be an achievement test because the types of questions they are presented with require them to memorize the materials and steps of the teacher-demonstrated experiments. The underlying problem relates to lack of clarity and adherence to the role of practical work in upper secondary science teaching and the commensurate underutilization of lab facilities.

The ministry needs to engage in more work directed at developing teachers' skills in designing and implementing assessment tools in tandem with developments in the teaching process. Ambusaidi and Al-Rashidi's (2009) study of the difficulties Omani science teachers face when conducting internal assessment found that these difficulties include teachers not having time to check students' work and projects at school, having too many items to fill in on student record sheets, and large numbers of students in each class. Teachers have proposed some recommendations to improve the assessment process. These include the ministry conducting workshops and training sessions to develop the skills teachers need to effectively implement both formative and assessment tools, and teachers sharing good practices through peer visits within and between the schools in each region. Because teachers need to know how to design questions that assess students' higher-order thinking skills and how to design and properly implement the practical examinations, the ministry should evaluate how the practical examination is conducted with a view to amending it so that it reflects the scientific activities conducted inside the classroom.

Continuing Professional Development Issues

Another challenge facing the Ministry of Education is whether or not teachers can or will translate what they learn from CPD programs into improved classroom practice. Unfortunately, some teachers attend many workshops and training programs but fail to reflect what they learn in their teaching. The ministry needs to develop ways of following up with teachers after they have attended professional development courses so that it can ensure the effort and the money spent on these programs are not being wasted. Any program that is well planned, implemented effectively, takes into account teachers' needs and weaknesses, and has a good follow-up mechanism should reflect positively on student achievement. Development of an overall framework within which CPD for science teachers can take place is therefore essential (Bell 2006).

An evaluation of the training program for new teachers showed these teachers wanting the program to have more of a practical emphasis, which has meant including examples of best practice from the field, particularly in areas such as classroom management (Ministry of Education 2013a). The evaluation of the CPD program for experienced teachers discussed above shows teachers needing more CPD in pedagogical applications. The Ministry of Education and Sultan Qaboos University are considering this requirement. The evaluation (in the form of a questionnaire) was done at the end of the program, and the results are being used to improve the second round of the program.

The two rounds of incentive programs for teachers have also been evaluated (Ministry of Education 2013a). According to some teachers, the program is very beneficial but the content is considerable and cannot be accommodated in the time allocated. Teachers also pointed out that they would gain greater benefit from the program if the examples provided were from the Omani education context rather than from the lecturers' national contexts.

Trends and Developments in Upper Secondary Science Teacher Quality

The Omani Ministry of Education has made a significant effort to improve science teaching in all grades, including Grades 11 and 12. It has introduced numerous initiatives as part of this effort. These include conferences and symposia conducted in order to benefit from the experiences of other countries in upper secondary education. The ministry is currently reviewing the science curricula of upper secondary schools with the aim of linking curricular content and practices to twenty-first-century skills and the labor market. This review is being carried out in collaboration with Sultan Qaboos University, the employment sector, and other members of society.

The review will inform development of a new curriculum framework, based on Omani economic needs and twenty-first-century competencies, for all subjects including science from Grade 1 to Grade 12. It is hoped that this framework will be comprehensive enough to guide curriculum developers, teachers, examination designers, and parents. The part of the framework pertaining to the upper secondary grades is expected to detail what students should know, understand, and be able to do, thus ensuring they have the skills and knowledge they need for study in higher education. There is a hope that the implementation of the new curricula will take place over the next few years.

The results of previous studies on Omani science teachers (Al-Balushi and Al-Rawahi 2011; Al-Harthi 2008; Ambusaidi and Al-Rashidi 2012) revealed some science teachers, male teachers especially, finding it difficult to implement best practice in areas of education such as assessment, classroom management, and inquiry-based learning. One way of helping resolve this issue would be for those teachers to sit an examination in both content and pedagogy before they are appointed as teachers and then participate in CPD programs. Recently, the Ministry of Education, in collaboration with Sultan Qaboos University, set up such an examination for new teachers. The examination includes items relating to subject matter, the curriculum, teaching methods, and psychology. Teachers who fail the examination are not allowed to teach until they pass it.

Another CPD initiative on the part of the ministry is its recent establishment of a specialist center for professional inservice training of teachers. The center's work focuses primarily on developing teachers' pedagogical content knowledge. Science teachers, for example, receive training in inquiry-based learning, active learning in science, and related matters. On completing this one-year program, teachers receive a certificate. The center uses three means of delivering its training content—face to face, online, and onsite.

In order to support the ministry's efforts to improve the quality of science teaching in Oman, the country's teacher education institutions need to reconsider their current science teacher education programs and try to develop them to be more effective. For example, the College of Education at Sultan Qaboos University is currently working toward gaining accreditation for its teacher education programs from the National Council for the Accreditation of Educator Preparation in the USA, and a number of changes have accordingly been made to those programs as an outcome of this initiative. It is worth mentioning that the university's science education program at Sultan Qaboos University was the first program in the college to receive accreditation. A number of changes have been made to those programs as an outcome of this initiative. For example, the practicum now takes place across one full semester compared to the previous arrangement, which was one day in the first semester and two days in the second semester. Student teachers can now take a course in research methodology and another in the ethics of teaching. For the science education program, the content knowledge part of the program has been aligned with National Science Teachers Association standards. Among the various other changes made is the introduction of a special course in health and safety. It is hoped that these changes will heighten the pedagogical skills of Grades 11 and 12 teachers.

Finally, in order to further develop science teaching at upper secondary schools in Oman, there is a need for more research on the impact on students of the issues raised in this chapter. Having put the spotlight on these issues, the science education unit at Sultan Qaboos University's College of Education is now encouraging its Master's and Doctoral students to tackle them in their research projects.

References

Al-Balushi, Sulaiman, and Nasser Al-Rawahi. 2011. "Investigating Omani Physical Education and Science Teachers' Beliefs in Cooperative Learning Using the Theory of Planned Behavior." *The Educational Journal (Kuwait University)* 101 (26): 285–322.

Al-Harthi, Aisha. 2008. "The Relationship between Science Teachers' Beliefs about the Use of Inquiry-Based Learning Strategy and Their Classroom Practices." Master's diss., Sultan Qaboos University, Sultanate of Oman.

Al-Rasbi, Nasser, Maryam Al-Nabhani, Abdullah Ambusaidi, R. Al-Fahidi, and M. Ibrahim. 2012. *Factors Affecting Students' Achievements in Grades Five and Ten in the Sultanate of Oman.* Muscat: Ministry of Education.

Ambusaidi, Abdullah, and Sulaiman Al-Balushi. 2015. "Science Education Research in the Sultanate of Oman: The Representation and Diversification of Socio-Cultural Factors and Contexts." *Science Education in the Arab Gulf States: Visions, Sociocultural Contexts and Challenges,* edited by Nasser Mansour and Saeed Al-Shamrani, 23–48. Rotterdam: Sense Publishers.

Ambusaidi, Abdullah, and Thuraiya Al-Rashidi. 2009. "Difficulties in Applying Formative Assessment in the Science Curriculum from Science Teachers' Points of View." *Journal of Educational and Psychological Sciences (University of Bahrain)* 10 (2): 147–66.

———. 2012. "Science Teachers' Attitudes towards Using Science Reading in the Classroom and Its Relations to Some Educational Variables." *Journal of University of Damascus for Educational and Psychological Sciences* 28 (2): 315–45.

Bell, D. 2006. "Continuing Professional Development: Enhancing Professional Expertise." In *ASE Guide to Secondary Science Education,* edited by Valerie Wood-Robinson, 40–8. Hatfield: Association for Science Education.

Ministry of Education. 1998. *The General Framework of Science and Math Curriculum for Basic Education.* Muscat: Author.

———. 2013a. *Professional Development and Development of Performance,* 3rd ed. Muscat: Author.

———. 2013b. *Professional Development Plan for 2013.* Muscat: Author.

———. 2014. *Learning Today for a Sustainable Future.* Muscat: Author.

Further Reading

Ambusaidi, Abdullah. 2000. "An Investigation into Fixed Response Questions in Science at Secondary and Tertiary Levels." PhD diss., University of Glasgow, UK.

———. 2010. *Promoting Innovation and Good Practices in ESD. Country Case Study: Sultanate of Oman.* Muscat: the Oman National Commission for Education, Culture, and Science, Ministry of Education.

Ambusaidi, Abdullah, and Ali Al-Shuaili. 2009. "Science Education Development in the Sultanate of Oman." In *The World of Science Education: Arab States,* edited by Saouma BouJaouda and Zoubeida R. Dagher, vol. 3, 205–19. Rotterdam: Sense Publishers.

Ministry of Education. 1998. *Assessment of Students' Learning in Science for Grades 5-10 Report.* Muscat: Author.

Ministry of Education and the World Bank. 2012. *Education in Oman: The Drive for Quality.* Muscat: Ministry of Education.

National Center for Statistics and Information. 2014. *Statistical Year Book 2013,* Issue 42, http://www.ncsi.gov.om.

11

Turkey

Muammer Çalik

Teaching Science in Turkey

Backdrop

The modern nation of Turkey was established in 1923 after the collapse of the Ottoman Empire. This country of 76.7 million people is divided into seven geographical regions across which climatic conditions, population density, and socioeconomic profiles differ markedly. The population is predominantly Muslim (98 percent) but includes ethnic and religious minorities such as Jews and Christians.

The Turkish education system came under central government control through the Law of Unification of Instruction in 1924, which brought all educational institutions under the auspices of the Ministry of National Education. The Basic Law of National Education was passed in 1973. It laid down essential goals of the reform strategy under the themes of the integrity of the education system, continuity in education, student selection, orientation and guidance in education, productivity in education, programs, instructional methods and materials, "education everywhere," and education and development (Binay 1982). Nearly all school education occurs in the government-run sector, the exceptions being schools run by ethnic minority groups. Education in public-sector schools is free of charge and is compulsory for students between the ages of 6 (Grade 1) and 17 (Grade 12). About 65 percent of children attend some form of early childhood education prior to Grade 1.

The compulsory school system comprises three levels: primary (Grades 1 to 4), lower secondary (Grade 5 to 8), and upper secondary (Grades 9 to 12). In Turkey, the transition from primary school to lower secondary within the public sector is determined by residential location. However, the transition from lower secondary to upper secondary involves sorting students according to their performance in the nationwide TEOG examination

(*Temel Öğretimden Ortaöğretime Geçiş Sınavı*) at the end of Grade 8. The TEOG comprises questions on Turkish language, mathematics, science, religion and ethics, modern Turkish history, principles of Atatürk (the founder of the modern Turkish state), and a foreign language such as English, French, or German. Destinations are the science secondary schools, social science secondary schools, Anatolian teacher training secondary schools, Anatolian secondary schools, Anatolian religious secondary schools, and vocational and technical secondary schools. The science secondary schools and social science secondary schools constitute the elite of the upper secondary sector and attract the academically strongest students.

School Science

The Scientific Commission for Development in Science Teaching (*Fen Öğretimini Geliştirme Bilimsel Komisyonu*) was set up in the late 1960s to translate and adapt foreign curricula (Physics Sciences Study Curriculum, CHEM Study, Biological Sciences Curriculum Study, School Mathematics Study Group), conduct inservice training for teachers, and conduct formative and summative evaluation of those curricula (Ayas, Çepni, and Akdeniz 1993; Turgut 1990). Many problem issues arose with regard to the curricula that had been translated and adapted, prompting policymakers to abolish them (Ayas et al. 1993). Following the abrupt demise of the Eastern bloc, particularly the breakup of the USSR, four new attempts were made (in 1992, 2000, 2005, and 2013) to reform the science curriculum in Turkey (Çalık 2014; Çalık and Ayas 2008). These attempts have all emphasized student-centered learning. A loan from the World Bank has been used for further curriculum development, building schools, and improving the resource base of existing schools as well as equipping teacher education institutions with new teaching technologies and laboratory equipment (Çalık and Ayas 2008).

The 2013 science curricula were developed and disseminated by the Scientific and Technological Research Council of Turkey (TÜBİTAK) acting under the authority of the Head Council of Education and Morality. Science at the primary school level involves three hours weekly in Grades 3 and 4 and is taught by generalist primary teachers. The curriculum is organized into four specific learning fields (creatures and life, matter and change, physical phenomena, and the world and universe). Science courses at the lower secondary level are taught by subject-specialist science teachers who have undergone an integrated teacher education program that includes physics, chemistry, and biology. Science is conducted four hours weekly in Grades 5 to 8 for the mandatory curricular components. Depending on student numbers (a minimum of ten being required for a class) and the availability of specialist teachers, such elective courses as science applications, environment and science, information technology, and mathematics applications may also be offered at this level.

Overall, the functions of upper secondary science education are to inculcate scientific literacy and to prepare students for higher (tertiary) education.

Students are allocated to one of three streams: science, social science, or one incorporating both of these strands. Science courses at the upper secondary level are taught by subject-specialist teachers in physics, chemistry, and biology. As well as specifying academic content, the 2013 physics, chemistry, and biology curricula make reference to the following subjects:

- Scientific literacy;
- The nature of science;
- Understanding scientific knowledge;
- Skills (scientific process skills and life skills such as informatics skills, group/team working, creativity and innovation, problem solving, awareness of responsibility, entrepreneurship, and communication);
- Science/technology/society/environment/economy interrelationships;
- Attitudes and values; and
- Psychomotor skills.

Now that a 12-year compulsory education cycle (4+4+4) has replaced the previous 8-year compulsory cycle, TÜBİTAK and the Ministry of National Education have needed to update upper secondary science curricula and produce new textbooks. However, the dearth of research-based evidence or needs analysis to support this educational move, the failure to take resource base into account, and of course the resultant large class sizes at the upper secondary level all militate against the success of these changes.

Because the number of students who graduate from secondary schools today vastly exceeds the current capacity of universities in Turkey, transition to tertiary education is highly dependent on performance in the national entrance examinations (*Yüksek Öğretime Geçiş Sınavıve Lisans Yerleştirme Sınavı*). This situation places upper secondary schools, especially the elite ones, in positions of high demand as bridges to competitive-entry tertiary education.

Teacher Academic and Professional Education and Training

As of 1998, upper secondary teachers were trained through a five-year teacher education program that consists of courses focused on both subject-matter knowledge and pedagogical content knowledge. An alternative route to teaching was a one-year postgraduate program of pedagogical content courses for science degree graduates in physics, chemistry, or biology. At the behest of faculties of science, the Higher Education Council (*Yüksek Öğretim Kurumu*) recently changed this system. In 2013, the council stopped admissions to the secondary science and mathematics education program delivered in faculties of education with the aim of redirecting students to faculties of science and thereby increasing the number of enrollments in science. However, the Council of State reversed this decision in 2014, and the Higher Education Council has accordingly reopened student admission to the secondary science and mathematics education program, which it has now reduced to a four-year program. The council announced as a rationale for this move the need to generate a competitive environment between faculties of education and faculties of science.

Today, the most common pathway to secondary teaching is a four-year teacher education program combining subject-matter and pedagogical courses. As an alternative pathway, undergraduate science students who still have to complete their subject degrees can attend weekend courses to attain a trained teacher certificate. This structure has, however, tended to lower the image of upper secondary teacher education by enticing lower-ability science students to enter teacher education. The move has also created disquiet among science educators about teachers' academic qualifications and training.

After graduating, newly qualified teachers who are willing to be employed in public-sector secondary schools sit a three-stage nationwide examination—the Public Personnel Selection Examination (*Kamu Personeli Seçme Sınavı*). The first stage of this examination gauges applicants' general knowledge (Turkish language, mathematics, history, geography, civics). The second stage tests applicants' knowledge of general educational principles and procedures (measurement and assessment, guidance, program development). The last one examines applicants' subject and pedagogical content knowledge.

The Directorate of Human Resources at the Ministry of National Education announces the number of vacant positions in secondary schools and appoints candidates with adequate examination scores as probationary secondary school teachers to those positions. These teachers become permanent staff at the end of their first year of experience, providing they receive favorable inspection reports. Those who receive unfavorable reports are relieved of their position.

Continuing Professional Development

The General Directorate of Teacher Education and Development has been delivering inservice professional development via seven regional inservice institutes (located in Ankara, Aksaray, Erzurum, Mersin, Rize, Yalova, and İstanbul) since 1960. The directorate organizes national and international seminars, symposia, and conferences on teacher education, carries out various professional development initiatives in collaboration with university faculties of education, and regularly conducts inservice education courses. It also regularly monitors teachers' performance and provides feedback, engages in research, and supports postgraduate education. It targets teacher professional competencies, initially introduced through preservice education, and prepares teachers for imminent changes in areas such as curriculum, philosophy, measurement, and assessment.

Since 2005, in response to the growing number of teachers in the public sector, the General Directorate of Teacher Education and Development has run distance education activities for inservice education. One such is the directorate's Constructivist Interactive In-Service Education Program. The directorate is planning to increase its level of collaboration with university departments of secondary science and mathematics education as part of its intention to make inservice education targeting the upper secondary level more effective. TÜBİTAK also provides some funding for inservice education, such as summer and winter schools, seminars, and scientific activities.

Science Teacher Professional Associations

A number of professional associations in Turkey support science education. On a general level, the Turkish Education Association organizes workshops and seminars for teachers (including upper secondary science teachers) and stakeholders and publishes research reports and the *Education and Science* journal. At the more specific level, the Turkish Science Education and Research Association, a community of teachers, researchers, and policymakers, is committed to improving science teaching and learning through research. The association takes as its mission providing good quality research findings to the educational community. It also organizes professional development activities in conjunction with various educational and scientific societies and publishes the *Science Teaching Journal*, which is practice oriented.

Every second year, the Turkish Chemical Society holds a national chemistry education conference in collaboration with the General Directorate of Teacher Education and Development and the universities. It also offers relevant professional development seminars in chemistry education. The society publishes chemistry education resources for chemistry teachers. The Turkish Physics Foundation undertakes similar activities for physics education.

Issues in Upper Secondary Science Teacher Quality

Teacher Supply

As noted earlier, the decrease in the perceived prestige of upper secondary teaching has brought about the situation of science teacher education programs attracting primarily lower-ability science students. At the same time, the General Directorate of Teacher Education and Development has drawn attention to the fact that few upper secondary science teachers will be needed in coming years. In excess of 8,000 upper secondary physics student teachers, 6,500 upper secondary chemistry student teachers, and 8,500 upper secondary biology student teachers took the third part of the Public Personnel Selection Examination (KPSS-ÖABT) in 2014. These numbers are large in relation to demand, and only small proportions have been recruited by public schools—about 8 percent of the physics and chemistry candidates and 11 percent of the biology candidates.

Despite this imbalance between supply and demand and despite the objections of deans of faculty of education and the General Directorate of Teacher

Education and Development, the Higher Education Council decided to offer, as an extra recruitment initiative, an intensive certification program (conducted over a series of weekends) for 60,000 undergraduate senior students (National Education Council 2010). This number included about 2,900 places for intending upper secondary physics teachers, 2,300 for chemistry, and 2,500 for biology. The council also advised of its intention to offer an additional 60,000 places the following year.

These numbers have created a situation that the Turkish educational community refers to as "unemployed teacher inflation." At the same time, the short-term alternative pathway to teaching is deemed unfair by various stakeholders because student teachers from faculties of education have to obtain almost twice the number of credits in pedagogical courses that students in faculties of science have to obtain, and yet both are eligible to apply for the same position if they have a high enough score in the Public Personnel Selection Examination. This paradox is taking its toll on the motivation and enthusiasm of student teachers from faculties of education.

Given the oversupply, it is unsurprising that teacher retention has not been a significant issue for policymakers in recent years. The need for upper secondary science teachers seems to have remained at a constant level, and an increase in retirement age is forcing teachers to continue working, so reducing the number of new teachers required. A commonly held view is that the global financial crisis made teachers postpone their retirement in order to make up for lost retirement savings. The age profile of the teaching force has steadily changed in favor of older age groups year by year. Given the number of student teachers graduating from faculties of education and science, the Ministry of National Education has no difficulty finding enough applicants when teachers do retire.

The number of teachers employed in public schools has increased slightly in recent years. However, as the number of teaching applicants increases, the quality of undergraduate entrants decreases. To address such concerns, TÜBİTAK began providing (in 2013) scholarship options for students achieving the best scores on the undergraduate placement examination (*Lisans Yerleştirme Sınavı*). Unfortunately, the "best" students have not taken up education; rather, these students aim for the elite universities and programs that lead to the most lucrative employment opportunities in the private and public sectors.

Another problem is emerging from the government's move to have all private tuition providers upgrade their institutions to private schools. In addition, the Ministry of National Education now subsidizes private school fees to make these schools more attractive to students and their parents and thus lessen the burden on the public schools. These developments call into question the potential fate of some 50,000 teachers (including almost 10,000 upper secondary science teachers) in the currently largely informal private tuition market (about 3,600 providers). The Ministry of National Education has responded by declaring its intention to recruit these teachers for the public education centers that prepare students to enter tertiary education. However, no one has explained in what way these public education centers will differ from the private tuition markets, except that they will be free of charge. Selecting these teachers through interview protocols without a nationwide examination raises several concerns regarding their assimilation into the planned system. Overall, the effects of these changes on other elements of the education system have not been well thought through, but they are being implemented nonetheless (Çalık 2014).

These various issues will eventually lead to a shortage of qualified teachers, particularly with regard to pedagogical competence. A Turkish saying seems fitting at this point: "The stew made from cheap meat will be tasteless (hard)."

Professional Education and Training Issues

The Ministry of National Education provides schools with textbooks, instructional materials such as CDs and DVDs, and Internet access. As a response to criticism that teachers become over-dependent on these materials (teacher guidebooks especially) when delivering the curriculum, the ministry decided to lessen teacher dependence by revising the curriculum and requiring teachers, as part of this change, to design and implement more of their own teaching materials. But old habits die hard, and most teachers, despite the revised curriculum, are still waiting for the new teacher guidebooks to arrive.

The revised curriculum also recommends that teachers use complementary measurement and assessment, thereby raising their need for professional development and training. And even though the upper secondary science curriculum, in accordance with the changes, requires its teachers to use complementary measurement and assessment as well as inquiry-based science pedagogy, the high profile of the national high-stakes examinations tends to dampen teachers' use of inquiry-based methods both outside and within school science laboratories. The examination-driven education system encourages rote learning and memorization of curricular content instead of comprehension and practical skills.

The revised upper secondary science curriculum also expects teachers to design assessment tasks and carry out moderation of the results, placing considerable demands on teachers' time and effort. For example, the topics "energy in industry and life," "chemistry is everywhere," "chemistry and energy," and "chemistry in our life" that were included in the 2013 upper secondary chemistry curriculum required teachers to enrich and update their own knowledge as well as generate resources and assessment materials. Because these units had previously not been taught at an advanced level, a demand for professional development emerged. However, inservice professional development has not been able to keep up with the numbers needing it. The Ministry of National Education is planning to address the problem by collaborating more closely with university faculties of education, but there has been no action on this matter to date. Both these problem areas indicate shortcomings in preservice education and training.

In 2011, the General Directorate of Teacher Education and Development released specific-domain competencies for upper secondary teachers. For example, the specific-domain competencies for upper secondary chemistry teachers include three principal fields:

- Content knowledge (e.g., evaluating the theories, laws, principles, hypotheses, and concepts pertaining to chemistry and relating chemistry to other courses and disciplines);
- Pedagogical content knowledge (e.g., evaluating the curriculum and analyzing students' pre-existing knowledge and learning difficulties); and
- Chemical literacy (Demirelli et al. 2011).

However, preservice upper secondary education still needs to catch up with these competencies and the demands of the 2013 curriculum, while teachers' competencies need to be improved in line with international moves toward twenty-first-century learning environments.

The Higher Education Council requires universities to revise their undergraduate programs in accordance with the Bologna Continuum Report, which stipulates that 25 percent of all undergraduate programs consist of elective courses. A four-year teacher education program (eight semesters) covers a total of 240 European Credit Transfer System (ECTS) credits: 180 for mandatory courses and 60 for elective courses. The first year of the program for upper secondary science teaching embraces such general compulsory courses as Turkish, mathematics, Atatürk's principles and history of the Turkish revolution, a foreign language (English, French, German), introduction to educational sciences and developmental psychology, basic information and communication technologies, and laboratory courses in physics, chemistry, and biology.

Student teachers attend specialized science courses from the second year of the program. For example, upper secondary chemistry student teachers take such compulsory chemistry courses as analytical chemistry, organic chemistry, physical chemistry, and inorganic chemistry. They can also choose from elective chemistry courses. These include environmental chemistry, industrial chemistry, stoichiometry, biochemistry, polymer chemistry, and food chemistry. Students are also enrolled in compulsory pedagogical content knowledge courses taught by chemistry educators. These include chemistry teaching I and II, measurement and assessment in chemistry education, instructional technologies and material design, learning/teaching theories and approaches, scientific research methods, school experience and teaching practicums as well as such compulsory general educational courses as classroom management and guidance (taught by general educators).

Chemistry student teachers furthermore have to select from such elective pedagogical content courses as curriculum analysis in chemistry education, misconceptions in chemistry education, complementary measurement and assessment in chemistry education, designing chemistry experiments, chemistry in our lives, chemistry/technology/society, and the history of chemistry. Student teachers generally attend faculties of science for the specialized science courses (such as at Karadeniz Technical University), although some faculties of education recruit specialized staff as science educators in preference to sending their students to the faculties of science.

The school experience and practicum requirements of Turkey's teacher education programs result in trainee teachers spending significant amounts of time each week in schools with teachers and teacher educators. Faculties of education work in partnership with schools to provide these experiences. The school experience course focuses on observation of teaching and classroom environments while the teaching practicum embraces teaching experiences (maximally six class hours in a 14-week term). Groups of trainee teachers (about six to seven per group) are assigned to teacher mentors (who receive a modest remuneration for this role). Although this partnership seems to be working in general terms, concerns have been raised over how it might evolve and improve. One of the most serious concerns is about the quality of mentors. The selection of mentors is solely at the discretion of the host schools. Good teachers are sometimes reluctant to be involved in student teacher training due to examination-related pressures as exerted by administrators, parents, and students. Finding schools willing to host student teachers is becoming increasingly difficult.

The Turkish National Committee of Teacher Education (*Öğretmen Yetiştirme Türk Millî Komitesi*), made up of representatives from the Higher Education Council, the Ministry of National Education, and faculties of education, was founded in 1997 and abolished in 2012, whereupon its mission was taken up by the Working Group for Teacher Education (*Öğretmen Yetiştirme Çalışma Grubu*). The main aspects of this mission are to enhance the quality of preservice education through collaboration with faculties of education; determine and implement national criteria for preservice education; develop quality control mechanisms to evaluate and improve the quality of teacher education; generate national standards for teachers with regard to knowledge, skills, and abilities; and devise an active role for schools with respect to preservice education (Working Group for Teacher Education 2014). However, indications are that this group is not very active or that policymakers do not take the group's views and decisions into account. It would be better if the Turkish National Committee of Teacher Education were reactivated as an independent entity instead of depending on the Higher Education Council and the Ministry of National Education.

Issues at the Chalkface

A significant challenge for upper secondary science teachers over recent years has been following and implementing the upper secondary science curricula. Despite these curricula having been updated a few years earlier, they were revised again in 2013 through the addition of new dimensions such as life skills, twenty-first-century skills, and integrated scientific process skills. These frequent changes in the curriculum make it hard for teachers to adapt, yet the implementation of each change is left mainly to teachers' own efforts due to the limited inservice preparation they receive prior to the changes taking effect. Teachers are expected to take active roles in "upper secondary science groups" as a form of self-directed professional development (under the euphemism "lesson study") aimed at translating the new curriculum into classroom practice. However, all that this endeavor seems to have done is increase teachers' workloads, despite rhetoric from politicians and officials claiming that teachers can well afford the time because of their "limited workload" of 15 classes a week.

Because the 2013 curricula require teachers to develop secondary science students' scientific process skills, teachers need sufficient laboratory equipment. But even when they do have sufficient equipment, they typically face the dilemma of whether to engage in the didactic teaching that upper secondary students need for passing the university entrance examination or to encourage the student-centered, inquiry-based learning that the curriculum calls for. The choice is probably made easier by the fact that secondary schools (and their teachers) are evaluated according to the number of their students who are admitted to the prestigious tertiary programs and/or universities. Most teachers focus on exam-oriented learning rather than on laboratory tasks, making laboratory tasks seem trivial or inferior.

A significant proportion of secondary science students graduate without seeing the inside of a school laboratory. The low level of laboratory use in upper secondary science education may also stem from a lack of suitable science technicians to help the teachers run laboratory tasks and maintain and prepare equipment; teachers do not have the time to set up the equipment on their own. A complicating factor is that teachers are generally expected to teach crowded classes. Statistical information (Ministry of National Education 2014) shows that the mean number of students in each class at the upper secondary level is about 39—almost double the acceptable level of 20 students per class. Such numbers place severe constraints not only on engagement in practical work but also on meeting the overall requirements of the 2013 upper secondary science curricula. All of that aside, practical skills and group investigations are not explicitly included in the formal assessment system, which accordingly sees teachers paying little attention to them anyway.

The 2013 upper secondary science curricula call upon teachers to design and administer complementary measurement and assessment tasks beyond the conventional item types such as multiple-choice questions, one-word answers, one-sentence definitions, and the labeling of diagrams. This added work requires teachers to spend significant time on complementary measurement and assessment as well as formative feedback. Changes in measurement and assessment have forced teachers to move from the traditional transmission teaching of content to such contemporary pedagogies as group work, problem solving, formative assessment, and peer- and self-assessment. However, in reality, upper secondary students are evaluated through examinations made up of multiple-choice items and/or open-ended questions. This adherence to the multiple-choice format stems largely from the Center for Measurement, Selection, and Placement's (*Ölçme Seçmeve Yerleştirme Merkezi*) responsibility for devising and conducting nationwide summative examinations.

The mismatch between the requirements of the curriculum and demands of the students and their parents has caused considerable angst for teachers. The lack of measures of curricular effectiveness and teaching methods results in assessments being used solely for filtering purposes rather than for gaining information about students' learning difficulties and deficiencies, and can lead to confrontations between teachers, slower learners, and their parents.

The two broad purposes of school science education are to provide "preprofessional education," the intention of which is to prepare students to enter tertiary education in science, and to help students meet "citizen-focused objectives." These objectives embrace scientific literacy, the interrelationships between science/technology/society/environment, and twenty-first-century skills (analytical and critical thinking, creativity and innovation, group/team working, entrepreneurship, and awareness of responsibility). The overarching idea behind facilitation of these skill sets is that they will prepare students for the complex role that science will play in their adult lives and for when they need to make decisions relating to issues arising out of the science-related discussions and debates they are likely to encounter. For instance, what they learn in school should enable them to verify via the Internet and/or related documents the reliability of controversial information and issues.

Teaching these citizen-focused objectives requires teachers to utilize pedagogies that align with these objectives. However, knowledge/content approaches continue to dominate pedagogical practice in upper secondary schools, thus presenting a barrier for students wanting to continue to study science for its own sake (Çalık 2014). Yet, ironically, despite the pervasive knowledge-centered focus in upper secondary school, universities still complain about students' lack of content knowledge and students' decreasing ability to cope with professional science programs.

Continuing Professional Development Issues

Inservice professional development provision is prevalent in Turkey, but the large number of teachers employed in public schools has brought about a need to employ some non-traditional alternative approaches such as distance education and closer collaboration with universities. Inservice interventions are short term (mostly one week), and their effectiveness has yet to be evaluated through follow-up studies. Experienced teachers who are keen on continuing their professional development through postgraduate programs have to compete for places with newly graduated teachers.

The postgraduate programs use the grade point average and scores on the Academic Personnel and Postgraduate Education Entrance Examination (*Akademik Personel ve Lisansüstü Eğitimi Giriş Sınavı*) to list and align applicants. The examination is made up of multiple-choice questions relating to Turkish language, mathematics, digital logic, and verbal logic, and even experienced teachers tend to have difficulties gaining enough marks on it to get into a postgraduate program. If they do manage to achieve this goal, they still tend to struggle with course schedules and English (the language of most educational literature). In 2009, the Ministry of National Education founded a National Education Academy to address these issues, but to date there has been little movement on them. In addition, these programs are mostly offered in the major cities, so teachers working outside them often have to move to those places to take up study. These various difficulties act as disincentives for teachers to continue their professional development and restrict access for potentially good teachers.

Through a project called FATİH, the Ministry of National Education has begun equipping every school with smartboards as well as laptop computers for distribution among the student body. However, preservice and inservice education have yet to become involved to the large-scale extent needed to prepare teachers to adapt to the new teaching/learning environment being created. Simply equipping schools with new technologies and not preparing teachers to use them effectively will markedly limit Turkey's ability to keep pace with current international trends in education (Çalık 2014).

Alongside the move to equip classes with data projectors, access to the Internet, and smartboards (FATIH), the General Directorate of Innovation and Educational Technologies has been updating and enriching Turkey's Education Informatics Network (*Eğitim Bilişim Ağı*) with videos, e-journals, e-books, e-documents, and so on. Teachers can access the network and take from it what they want for their classroom practice. They can also share their ideas through the network's social site. However, to be truly useful, the network's functional framework needs to be more interactive rather than simply present curriculum-related documents.

Trends and Developments in Upper Secondary Science Teacher Quality

In Turkey, long-term trends and/or developments are not the norm because the system tends to be driven by short-term policies. Upper secondary science teacher quality indicators change from year to year. For instance, as noted earlier, due to quality concerns, the Higher Education Council abolished the short-term weekend certificate program, brought in distance teacher education programs, and then opened these up to 60,000 positions in the same year. Erratic decisions are run of the mill in teacher education. For example, despite the Ministry of National Education insisting that its technological pedagogical content knowledge (TPACK) program will replace short-term pedagogical certificate programs, the latter are set to continue. Such decisions show that teacher *quality* is not as important as teacher *quantity*. Quantity-focused decisions in upper secondary science teacher education have the potential to badly threaten future educational outcomes.

Turkey needs genuine long-term teacher education policies, especially for upper secondary science teacher education. The two-pronged competitive student teacher recruitment system seems not to be working in terms of enhancing the quality of teacher education. Turkey has 184 faculties of science in operation, but only 16 university faculties of education, including the Department of Secondary Science and Mathematics Teacher Education. The quality of upper secondary science teachers is usually attributed to the faculties of education even though, as students, these teachers will have spent much more time in subject-matter courses taught in the science faculties. This claim is used as justification for the short-term weekend certificate program (pedagogical formation).

Interestingly, Safran et al. (2013) provide confirmation of the notion that the education faculties do seem to positively influence teacher quality. They reported that upper secondary science student teachers from university faculties of education presented significantly better scores in the third part of the Public Personnel Selection Examination (KPSS-ÖABT) than did those students from faculties of science. Safran et al.'s (2013) study seems to refute the principal argument in favor of the short-term certificate program. However, because research-based reforms are not common in Turkish teacher education, policymakers and the Higher Education Council seem to have paid little if any attention to this report.

Research-based reforms are also badly needed in Turkish science education if teacher education is to be improved in reality and if inquiry-based teaching methodologies are to become common practice in schools. Coordination among university faculties of education, the General Directorate of Teacher Training, and the Higher Education Council also need to be functionally enhanced with a view to producing long-term educational policies in Turkey and professional unity of purpose and direction vis à vis examination-oriented education.

Only 2 percent of science secondary school students select a faculty of science for their future career pathways, while only 20 percent of Anatolian teacher training secondary schools choose a faculty of education (TEDMEM 2014). Ways of improving the public image of these faculties should be investigated carefully. Despite TÜBİTAK offering very good scholarships to students wanting to study in any department within a faculty of science, the faculties of science remain undersubscribed. The Higher Education Council needs to encourage the faculties of science to revise their standards through university-industry relationships.

To encourage teachers to continue on to postgraduate programs (Master's degrees, PhD/EdD), the Ministry of National Education is planning to provide incentives for teachers, such as giving teachers extra promotion points if they are prepared to move to major cities where these programs typically take place. Unfortunately, the ministry has still not sufficiently distinguished between teachers on the basis of qualifications; salary scales remain the same across them, thus limiting teachers' incentive to become better qualified. And, as it turns out, almost every teacher with a PhD wants to transfer to a university and take up an academic position. Policymakers therefore need to rethink how to make effective use of these highly qualified teachers in the professional development of other teachers.

Admittedly, as described earlier in this chapter, the Ministry of National Education did introduce the ranks of expert teacher and lead teacher in 2005 and offered teachers willing to take up these roles extra monthly pay. However, this so-called career system lasted only until the end of 2006 because of the large number of teachers complaining about it being akin to a caste system. Nevertheless, recent trends indicate that such a system is still needed to encourage better-qualified teachers to work within their own schools in order to lift standards of practice and improve achievement. Such teachers could serve as role models for teacher professional development not only within their own schools but also the other schools in the vicinity.

Turkish student achievement in international measures of educational achievement has not been at a satisfactory level (it has been below the OECD averages), although student scores have improved over time. That said, the validity and reliability of these measures as true indicators of student achievement have been critically questioned. There is a need for studies that monitor the relationship between student achievement and teacher quality.

Although all upper secondary science curricula recommend the use of complementary measurement and assessment, our teacher selection and placement system remains traditional. The reason commonly given for this is the huge number of applicants. Alternative ways to recruit the best applicants should be looked into. Private schools often select teachers through strict interview protocols as well as microteaching demonstrations. The Ministry of National Education has considered implementing the same procedure for teacher employment in public-sector schools and/or delegating this presently centralized role to authorized people in the localities. But concerns have been raised about following standardized objective criteria (including grade point average scores) given the number of applicants. Another development in upper secondary education is the rotation of teachers whereby teachers transfer to a different school every eight years. The intention underlying the rotation is to ensure teachers are not left languishing in suboptimal teaching environments and to give students opportunities to encounter good teachers.

Currently, public-sector teachers may complete their entire teaching term without any inservice education and/or taking any examination leading to accreditation of their professional development. Ideally, teachers should regularly be acquainted with trends and developments in order to enhance their understanding of the curriculum and to properly implement it or to track contemporary learning approaches. Incentives encouraging teachers to take up professional development, such as giving credit points for different types of teacher appointments and promotions and/or revisiting the utility of having expert teachers and lead teachers need to be evaluated anew. School summer and winter breaks could be effectively used to bring up to speed those teachers who need to upgrade their skills and qualifications.

References

Ayas, A., S. Çepni, and A. R. Akdeniz. 1993. "Development of the Turkish Secondary Science Curriculum." *Science Education* 77 (4): 433–40. doi: 10.1002/sce.3730770406

Binay, H. 1982. *Turkey: Sociocultural Information*. Strasbourg: Council for Cultural Cooperation.

Çalık, Muammer. 2014. "Turkey." In *Issues in Upper Secondary Science Education: Comparative Perspectives*, edited by Barend Vlaardingerbroek and Neil Taylor, 229–41. New York: Palgrave Macmillan.

Çalık, Muammer, and Alipaşa Ayas. 2008. "A Critical Review of the Development of the Turkish Science Curriculum." In *Science Education in Context: An International Examination of the Influence of Context on Science Curricula Development and Implementation*, edited by Richard K. Coll and Neil Taylor, 161–74. Rotterdam: Sense Publishers.

Demirelli, Havva, Nejla Yürük, Nusret Kavak, Merih Ceritoğlu, Mithat Büyükhan, and Elif Özdemir. 2011. *Kimya öğretmenI özel alan yeterlikleri* [Specific-domain competencies for upper secondary chemistry teachers]. Ankara: General Directorate General of Teacher Training and Deployment, http://otmg.meb.gov.tr/alan_kimya _ortaogretim.html (in Turkish).

Ministry of National Education. 2014. *National Education Statistics: Formal Education 2013/14*. Ankara: Author, http://sgb.meb.gov.tr/istatistik/meb_istatistikleri_orgun _egitim_2013_2014.pdf.

National Education Council. 2010. Onsekizinci *Milli Eğitim Şurası Kararları* [18th National Education Council Decisions], http://ttkb.meb.gov.tr/meb_iys_dosya lar/2014_10/02113646_18_sura.pdf; http://www.meb.gov.tr/duyurular/duyuru lar2010/ttkb/18sura_kararlari_tamami.pdf (in Turkish).

Safran, Mustafa, Adnan Kan, Mutlu Tahsin Üstündağ, Togay Seçkin Birbudak, and Osman Yıldırım. 2013. "An Investigation of KPSS 2013 Results in Terms of Candidate Teachers' Fields." *Education and Science* 39 (171): 13–25.

TEDMEM. 2014. *Öğretmen İstihdam Politikaları* [Teacher employment policies]. Ankara: Author, http://www.tedmem.org/haberler/2014/02/17/ogretmen_istih dam _politikalari_sorunlar_ve_guncel_tartismalar_paneli.html (in Turkish).

Turgut, M. F. 1990. Türkiye'de fen ve matematik programlarını yenileme çalışmaları [Revision of science and mathematics curricula in Turkey]. *Hacettepe Üniversitesi Eğitim Fakültesi Dergisi* 5: 1–14.

Working Group for Teacher Education. 2014. Öğretmen eğitimi çalışma grubu [Working group for teacher education]. Accessed from https://www.yok.gov.tr/docume nts/10279/49665/%C3%96%C4%9Fretmen+Yeti%C5%9Ftirme+%C3%87al%C4% B1%C5%9Fma+Grubu/4054fef9-ffc8-41a7-a909-ac8c5188de66.

Further Reading

Azar, Ali. 2011. "Quality or Quantity: A Statement for Teacher Training in Turkey." *Journal of Higher Education and Science* 1 (1): 36–8.

Bahar, Mehmet, Ali Murat Sünbül, and Muhittin Çalışkan. 2013. *Ortaöğretime öğretmen yetiştirme: Mevcut durum-sorunlar-öneriler* [Teacher Education for Secondary School: The current situation, problems and advice]. Eğitim Fakülteleri Dekanlar Konseyi Çalıştayı [Council of Deans of Faculties of Education Workshop] November 22–23 2013, Konya, Turkey (in Turkish).

Eğitim İzleme Raporu [Education monitoring report]. 2013. http://erg.sabanciuniv
.edu/sites/erg.sabanciuniv.edu/files/EIR2013.web.pdf (in Turkish).

Özoğlu, Murat. 2010. *Türkiye'de öğretmen yetiştirme sisteminin sorunlar* [Problems of
the teacher education system in Turkey: An analysis] Ankara: Ekonomive Toplum
Araştırmaları [Foundation for Political, Economic, and Social Research].

Ünver, Gulsen, Nilay T. Bümen, and Makbule Başbay. 2010. "The Effectiveness of a
Secondary Teacher Education Graduate Program According to Administrators,
Faculty Members and Students." *Educational Sciences: Theory & Practice* 10 (3):
1807–24.

12

Malaysia

Noraini Binti Idris

Teaching Science in Malaysia

Backdrop

Malaya gained independence from Britain in 1957, and the Federation of Malaysia was formed in 1963. The education system is centralized, with school funding, teacher education, curricula, and national school-leaving examinations administered by the federal Ministry of Education, with state education and district education departments given limited powers. In the Malaysian education system, primary school covers six years and is followed by three years of lower secondary school and two years of upper secondary school. Those who pass the Malaysia Certificate of Education or MCE (equivalent to British O-Levels) at the end of Form 5 (the fifth year of secondary schooling) proceed to sixth form, which occurs over two years ("lower sixth" and "upper sixth"), after which they sit the Higher School Certificate or HSC (equivalent to British A-Levels) to enable them to enter university. Alternatively, after attaining the MCE, students can choose to undertake certificates, diplomas, or other technical qualifications in the private colleges and institutes. The technical and vocational stream is the stream within the mainstream school system in Malaysia that accepts students after lower secondary, and there are separate upper secondary schools.

School Science

The Malaysian national science education philosophy states that "in consonance with the National Education Philosophy, science education in Malaysia nurtures a science and technology culture by focusing on the development of individuals who are competitive, dynamic, robust and resilient and able to master scientific knowledge and technological competency" (Ministry of Education 2006 ix).

Science teaching in Malaysia has undergone many changes during the country's efforts to make it relevant to changing times and to fulfil national objectives in line with national and global developments. Science teaching in this country began during the British colonial era. At the time, the education system was elitist in nature; only a small minority of students proceeded to secondary school to follow formal science studies. In 1939, a committee for the Straits Settlements and Federated Malay States formulated the objectives of science education and produced a science syllabus. At that time, only one secondary school in the Federated Malay States offered a science course, which was taught by teachers from overseas. Students used foreign textbooks and sat the Cambridge examination as their peers in England did.

This initiative was short-lived because of the outbreak of World War II. When the war ended, the subject "general science," based on the British syllabus of the time, was made part of the curriculum in all secondary schools. After Malaya gained independence in 1957, this policy was revised in appreciation of the important role that science education plays in economic and social development. Since implementation of the Second Malaysia Plan (RMK-2) in 1968, the government has emphasized its commitment to science and technology education. The transformation of science education in the wake of the Sputnik satellite launch was also felt in Malaysia.

The Curriculum Development Centre in Malaysia, created in 1973, aimed to monitor all aspects of the use and adaptation of Malaysia's curricula, including science. The importance attributed to science education was evident in the Razak Committee Education Report (1956), Rahman Talib Education Committee Report (1960), and the Cabinet Education Report (1979). Changes to the Malaysian school science curriculum occurring over subsequent years included implementation of the Special Project in Primary School Science, the Integrated Science Curriculum for lower secondary schools, and the Modern Pure Science Curriculum (biology, physics, chemistry, and integrated science) for secondary schools, along with the New Primary School Curriculum (KBSR) and the Integrated Secondary School Curriculum (KBSM).

The traditional secondary school curriculum implemented in 1950 into the 1960s was found to have several weaknesses, one of which was that it did not portray science in either content or approach as practiced by scientists. To resolve this problem, the Ministry of Education adopted innovative curricula from Western countries and introduced them in 1969. Among the changes implemented was the introduction of "modern pure science" for upper secondary schools. Modern pure science was made up of biology, chemistry, and physics subjects and was piloted in ten schools. The syllabus was adapted from the Nuffield O-Level syllabi. One of the objectives of teaching and learning science at this time was to expand students' understanding of scientific concepts and the use of science in real-life situations. The success of the pilot project led to the pure science curriculum being extended to all schools in Malaysia. However, what appeared to be a positive effort soon showed its limitations; the curriculum was found to lack connections between science, humanity, and society. Furthermore, the majority of teachers felt that the modern science curriculum was too focused on process, such that it neglected the fundamentals students needed to master as preparation for higher education.

A wave of education reform overwhelmed the nation once again in 1988, which marked the launch of the Integrated Secondary School Curriculum (KBSM). In alignment with the principles underlying the development of scientific knowledge and skills, the teaching and learning of science also stressed social values and positive attitudes toward science. Values and practices were closely related to the notions of using scientific knowledge and applying technology in ways that aided decision making on matters relating to human well-being and environmental balance.

The OECD's large-scale Programme for International Student Assessment (PISA) is a worldwide iterative study that examines the scholastic performance of 15-year-old students in mathematics, science, and reading. According to the 2012 PISA report (OECD 2012), Malaysia emerged as number 52 out of 65 countries. For scientific literacy, Malaysian students scored an average of 420 achievement scale points, well below the OECD average score of 501 points; this represented a decline from the 422 achieved in 2009 (OECD 2009). The performance of Malaysian students on the Trends in International Mathematics and Science Study (TIMSS), another large-scale study of student achievement conducted by the International Association for the Evaluation of Educational Achievement, has lagged behind that of students in other Asian, especially Southeast Asian, countries.

The main aim of the Malaysian primary science curriculum is to lay the foundation for building a society that is culturally, scientifically, and technologically dynamic and progressive. Science continues to be offered as a core subject to all students at the lower secondary level. At the upper secondary level, students can choose from science electives (biology, chemistry, physics, and additional science) in addition to core science.

Teacher Academic and Professional Education and Training

Academic entry qualifications are the same for lower and upper secondary teachers in Malaysia. All secondary teachers must have either a Bachelor's degree in education or a post-graduate diploma in education/teaching plus a Bachelor's degree in a teaching subject. Although Malaysia lacks a licensing system for public school teachers, candidates must pass a competitive screening process before acceptance into preservice teacher training programs. Their grades on their MCE or HSC have to be no lower than a specified level, and they need to pass the Malaysian Teacher Selection Test and an interview conducted by the Education Service Commission. Those who do enter and then complete the program are not appointed as permanent teachers in government service unless they achieve certain cumulative grades during their training and successfully pass another interview. Private school teachers must obtain a teaching license before starting teaching. This requirement is regulated by the Private Education Division of the Ministry of Education through each educational institution.

The Ministry of Education's Teacher Education Department is responsible for teacher training in cooperation with teacher training institutions (TTIs) and local public higher education institutions (HEIs) under the Ministry of Education. Candidates selected for preservice teacher training attend one of the 27 TTIs, which are strategically located in all 14 states of Malaysia. The TTIs provide inservice as well as preservice courses. Each institute can accommodate about 800 to 1,000 students. In the past, the TTIs were referred to as teacher training colleges. When the teacher training programs were upgraded from diploma to degree level, the status of the colleges was raised to that of higher education institutes. Since 2006, they have been referred to as teacher training institutes.

The majority of primary school teachers obtain their qualification from TTIs. Although both TTIs and HEIs have Bachelor degree programs, the four-year BEd program offered at HEIs is the most common choice for candidates seeking a secondary teacher qualification. Both types of institution offer post-graduate diploma programs for those with Bachelor's degrees in a specialized area or in a school subject. It takes one year to complete the full-time programs and 18 months to complete the school-based programs. The latter are intended for uncertified teachers in temporary service. Student teachers in preservice training programs are also required to undergo compulsory practicum training, the duration of which depends on the program and the institution (12 weeks for a BEd program in HEIs, 24 weeks in TTIs, and 12 weeks for students in postgraduate diploma programs).

Malaysia, unlike some other countries where teachers need to renew their teaching license after a certain period, does not have a relicensing system in place, although calls have been made to change this approach so that teachers can be accredited by an independent third party, as suggested by Mohd Salleh, Sulaiman, and Fredericksen (2014). Such a system would be hard to implement, though, given Malaysia's current centralized teacher education system. However, if it were adopted, having an independent third party evaluate teacher quality would encourage Malaysian teachers, including science teachers, to be responsible for their own learning and professional development, hence improving the quality of secondary school science teachers.

Continuing Professional Development

Inservice teachers in Malaysia wanting to upgrade their qualifications and/or engage in continuing professional development (CPD) have several options. The Teacher Education Department in collaboration with TTIs and HEIs is responsible for providing such training. In principle, all teachers are required to attend CPD courses for a minimum of seven days per year given it is a prime focus in the new teacher remuneration system. However, the total number of hours given over to CPD depends on the needs of individual teachers and the curriculum. The government also encourages teachers to upgrade their qualifications by providing them with scholarships to support the costs of their study.

Avenues for CPD among science teachers include courses at local universities and regional centers such as the Regional Centre for Education in Science and Mathematics (RECSAM). Launched in 1967 under the auspices of the Southeast Asian Ministers of Education Organization (SEAMEO), RECSAM provides courses on science and mathematics for Malaysian teachers as well as for teachers from the Southeast Asia region.

Science Teacher Professional Associations

Science teachers in Malaysia belong to the National Union of the Teaching Profession. However, unlike English teachers who have an active chapter called MELTA (Malaysian English Language Teaching Association), science teachers, probably because of their relatively low numbers, are less well organized. In Malaysia, science teachers can voice their concerns through *panitia* (school-level committees) or through popular science blogging sites.

The Ministry of Education encourages teachers to use the Frog Virtual Learning Environment, an online forum where teachers can exchange ideas, teaching resources, and air their views. Through its 1BestariNet project, FrogAsia is bringing high-speed 4G Internet access to the Malaysian education system, thereby promoting a world-class learning platform and access to teaching and learning resources and technology for every student, teacher, and parent in Malaysia. The ministry wants to bring the entire education community together on a single converged network designed specifically to meet teaching and learning needs.

Issues in Upper Secondary Science Teacher Quality

Studies specific to Malaysian upper secondary science teacher quality are rare. Consequently, the studies consulted for this chapter cover teacher education in general or mathematics and science teachers as a group, since many local academic studies focus on mathematics and science teachers in combination.

Teacher Supply

In Malaysia, teacher shortages most commonly occur in key areas such as Malay language, English language, mathematics, and science. Having stated its policy of making teaching an occupation of choice, Malaysia's government is striving to empower teachers and improve their working conditions. These endeavors cut across all subjects and levels of the education system and so include upper secondary science teachers.

According to the World Bank, the learning outcomes of Malaysian students on TIMSS were above the international average between 1999 and 2003 but declined sharply in 2007 and further again in 2011 ("World Bank" 2013). The World Bank advised Malaysia that if it was serious about halting the decline it needed to prioritize teacher quality over quantity. The bank noted in particular that the sharpest fall in education standards had coincided with an aggressively expanded recruitment program for educators. Between 2004 and 2013, Malaysia's teacher population increased by 30 percent. The number of trainees enrolled in teacher training institutes ballooned from 37,439 in 2011 to 46,491 in 2013 ("World Bank" 2013). The bank also noted as an associated problem the low academic standards of those seeking to become teachers. For example, 93 percent of applicants for the Bachelor of Education program lacked the requisite academic qualifications (i.e., three distinctions or more at the Malaysian Certificate of Education level) while 70 percent of those actually offered a place in the program also fell into this category. Only 3 percent of offers went to applicants considered high performers. The number of qualified teachers in Malaysia does mean, however, that the country presently has no difficulty staffing rural schools or replacing teachers who retire or leave the profession.

Professional Education and Training Issues

The level of cognitive mastery among preservice teachers studying in higher education institutions is below satisfactory. For example, Misbah and Shaharom (2008) found that 41 preservice teachers at Universiti Teknologi Malaysia (UTM) did not understand or know the communication and research skills needed to conduct a survey made up of a set of questionnaires. Mohamed Isa (2001) reported on 113 preservice teachers in Perlis who could not identify variables and could not set and use hypotheses.

Lazim et al. (2009) looked at the pedagogical skills of 115 final-year (science) teacher trainees at a Malaysian university and found that the trainees' overall competency in terms of their pedagogical as well as their content mastery was at a high level. However, the researchers also found that the trainees were generally unable to carry out science experiments without referring to the practical workbook and that their ability to construct science questions was poor. Lazim and colleagues recommended that the university determine why these teachers had difficulty carrying out science experiments and designing science questions and that it also identify solutions to these problems.

Tan and Chin (2001) found that teachers were highly adept at drawing inferences, mainly because of the effective teaching of their lecturers and sufficient exposure to inference skills while at university. In contrast to the findings of the above-mentioned study by Mohamed Isa (2001), Lazim et al. (2009) found that teacher trainees were well able to identify scientific variables and draw hypotheses. We can infer from these findings that science teacher quality in Malaysia improved between 2001 and 2009, the dates when Mohamed Isa and Lazim et al. carried out their respective studies. During the interval between these years, lecturers and curriculum designers paid serious attention to the types of issue Mohamed Isa identified and then worked hard to improve the quality of teacher education programs.

Issues at the Chalkface

One of the main issues in science education in Malaysia is the lack of exposure that students have to practical work in primary school, which leads to problems in secondary school science. During her case study on the alignment between science curriculum guidelines and actual teacher practice in a primary school, Selvaranee (2013) found that teachers were still practicing teacher-centered teaching instead of encouraging students to explore and investigate.

A major transformation occurred in 2002 following the introduction of the policy of Teaching and Learning Mathematics and Science in the English language. This policy, known by the Malay acronym PPSMI (*Pelaksanaan Pengajaran dan Pembelajaran Sains dan Matematik Dalam Bahasa Inggeris*) was to be implemented from 2003 in stages in all primary and secondary schools, national schools, national-type schools using minority vernacular languages as the medium of instruction, secondary schools, and Form 6 schools. However, some problems emerged during implementation (Alwis 2005; Idris et al. 2007; Ramlan 2009). Alwis (2005), for example, found that teachers had low mastery of English language for teaching mathematics and science, while Ramlan (2009) found from her qualitative study that teachers working under the PPSMI were code-switching between English and Bahasa Malaysia during science classes.

Language of instruction tends to be a problem in science education in Malaysia, especially in rural areas, where exposure to English outside the classroom is extremely limited as compared to the urban areas, where exposure to spoken English is more likely. Although PPSMI was abolished in 2012, many teachers are still using the teaching courseware developed for it by the Malaysian Ministry of Education. Textbooks, workbooks, and computer software continue to be popular as instructional resources at Form 2 level.

TIMSS 2011 revealed that the science results for Malaysian students had dropped drastically when compared to the results of TIMSS 2007. Critics in Malaysia cited this performance as evidence of low quality among new science teachers. Although the samples of students who participate in TIMSS are not drawn from the upper secondary level, the findings of the study are seen as an indicator of the quality of students making the transition to upper secondary.

The TIMSS 2011 international report (Martin et al. 2011) also revealed that Malaysian science teachers were more didactic than many of their international counterparts when teaching science. Students were mostly taught with textbooks as the basis for instruction. Less than 50 percent of students had access to workbooks or worksheets, science equipment and materials, and computer software as a basis of their learning. Malaysian teachers are influenced by the Confucian education culture, where most science lessons are centered on learning from textbooks and memorizing facts (Ling and Mohd Saat 2013).

Mohamad Nasri et al. (2010) researched the problems of science teachers in Malaysia. Their qualitative study showed that these teachers were facing many problems and that the problems could be grouped into three clusters: pedagogical, professional developmental (a lack of support by schools, government, and non-government agencies), and facilities (school science laboratories not well equipped, thus discouraging teachers from conducting active teaching and learning processes).

Research on Malaysian science teaching in the 1990s repeatedly documented the prevalence of traditional teaching techniques, particularly lecturing and note copying. Anuar Zaini et al. (2003) also found teachers using didactic teaching styles to complete the syllabus. These teachers were drilling students to memorize copious quantities of facts in the classroom in preparation for national examinations. In his study featuring 347 teachers in 123 schools, Samsuddin Jalil (1996) found that only 56 percent of these teachers understood what science process skills mean, only 45 percent had attained competence in manipulative skills, and just 30 percent comprehended what critical and creative thinking meant. Another Penang Education Department study showed that the 16 teachers who participated in it had weak science process skills, such as putting forward a hypothesis, engaging in experimentation, and making inferences (Penang Education Department 1996).

As part of their examination of Malaysian science education, Bevins et al. (2001) surveyed teachers in order to identify their views of science education provision in Malaysia. Bevins and colleagues reported that teachers saw Ministry of Education strategies for developing science education as incoherent and piecemeal, with this situation a product of the low priority awarded to science compared to arts, medicine, and law (and, indeed, few graduates enter science teaching as a preferred career option). The transition from primary to secondary school science was also seen as a challenge because of the difference in focus at each level. Although primary science is now attempting to develop investigative science and associated skills, such as hypothesizing and experimentation, secondary science still has a didactic emphasis on knowledge transfer and does not continue the emphasis on investigative science skills being encouraged at the primary school level.

In this context, the teachers in Bevins and colleagues' (2001) study also identified the need for considerably less focus on recipe-based experiments, more research-based investigations, and more student-centered approaches in order to enthuse and inspire students. Teachers wanted to see student-centered pedagogical approaches in classrooms because of these being better than didactic methods at creating student interest. They furthermore wanted stronger teacher–student relationships brought about by effective classroom management, creative teaching, problem solving, and assessment. More prosaic concerns included large class sizes (often 40 to 60 at the secondary level), a lack of resources, including inadequate laboratory equipment, outdated textbooks, and time constraints. The teachers also commented on the prevalence of female teachers in science education (not surprising given that teaching has traditionally been seen as a female profession in Malaysia). Overall, teachers saw students as exhibiting a low interest in science acculturated in the knowledge transfer approach. Students were continuing to demand information as part of "cramming" for the university entrance examination. There was a perceived need to inject fun and relevance into science teaching and to utilize varied approaches.

Continuing Professional Development Issues

In their study on Malaysian mathematics and science teacher professional development, Mukundan, Nimehchisalem, and Hajimohammadi (2011) found that while teachers expressed a high need for CPD activities, most of them regarded the programs and courses they had experienced as ineffective. Teachers also tended to say that they preferred release time to financial support and advancement on their pay scale as the incentive for participating in teacher professional development. One of the researchers' conclusions was that less experienced teachers were more in need of professional development activities than their highly experienced counterparts.

The TIMSS 2011 findings (Martin et al. 2011) found that even though Malaysian science teachers were using computers in their science instruction, they were seldom using this technology in a manner designed to enhance students' cognitive skills. Essentially, Malaysian science teachers use computer software didactically; as a result, students have weak process and reasoning skills relative to science subject content. This issue needs to be addressed in science teacher education programs both at preservice and inservice level. In addition, mechanisms need to be established that encourage teachers across schools to share examples of successful or unsuccessful ICT-mediated lessons. Such sessions could provide teachers with ideas on how to conduct computer-based lessons and would also let teachers see that if their colleagues can use ICT effectively, they can too. Teachers need to be trained in using ICT in constructivist ways in order to provide a flexible learner-directed workspace. This type of teaching is especially important for teaching and learning science topics in the new primary school curriculum (KSSR) and also the secondary school curriculum (KSSM), which is still in the development stage (Ling and Mohd Saat 2013).

The Ministry of Education is trying to incentivize teachers in this direction by, for example, providing scholarships for deserving candidates to pursue continuing education. In addition to the TTIs and HEIs, different departments and divisions within the ministry offer professional development courses in different areas. The Curriculum Development Center, for example, offers courses to help teachers accommodate new curriculum changes, while the Educational Technology Division provides courses on ICT use. Additional to these courses are the Aminuddin Baki Institute's school-management courses and the state education departments' and district education offices' provision of training focused on teaching for optimal learning and in response to curriculum reform.

Some of Malaysia's universities plan and organize short-term courses and seminars for inservice teachers. These courses typically seek to improve teachers' professional and personal development in general and their pedagogical skills and subject-based knowledge in particular. The School of Educational Studies of the Science University of Malaya also conducts workshops, seminars, and short courses for preservice teachers. The Faculty of Educational Studies at the Universiti Putra Malaysia organizes offshore and distance education programs for busy teachers who are unable to attend campus-based academic programs. Almost all faculties/schools of education in the country's 13 universities have organized national and/or international conferences, during which some attending inservice teachers present papers. Participation is regarded as engagement in professional development for teachers in Malaysia.

Trends and Developments in Upper Secondary Science Teacher Quality

In 2011, a policy titled Empowering Bahasa Malaysia and Strengthening English Language (MBMMBI) replaced the PPSMI, and science teaching reverted to the use of Malay language as the medium of instruction. However, a "soft landing" approach was taken to prevent difficulties for students who had already undergone PPSMI in their lower secondary years. They were able to continue studying science and mathematics in English until their major examinations. The MBMMBI policy has been implemented in stages in primary and secondary schools, with teaching of science and mathematics now conducted in Bahasa Malaysia. However, teaching of English as a second language has been strengthened at both teacher education and classroom level.

Implementation of the Standard Curriculum for Primary Schools (KSSR) and Standard Curriculum for Secondary Schools (KSSM) will hopefully further advance science teaching and learning. KSSR was introduced to Year 1 students in 2011 and by 2016 will have fully replaced the KBSR. The KSSM will begin with Form 1 (Grade 7) in 2017.

Upper secondary school science teacher quality in Malaysia should be improving given the emphasis on enhancing overall teacher quality in Malaysia as announced in the *Malaysia Education Blueprint 2013–2025* (Ministry of Education 2013a). The blueprint also notes the effort needed to strengthen teacher professional development in Malaysia. This effort naturally pertains to upper secondary science teachers as well. Another development of note is that of initiatives to enhance technology use in secondary education, especially in mathematics and science teaching.

Ensuring Malaysian students have the higher-order thinking skills they will need in the competitive globalized world economy relies very much on teacher education directed toward pedagogy that facilitates creativity and innovativeness among students. As part of its economic transformation program, the government is seeking to overcome deficiencies in the supply of quality English, Bahasa Malaysia, mathematics, and science teachers by giving the private sector a greater role in the preservice training of teachers. Inservice training of teachers will also be liberalized by allowing the private sector to conduct classes or hold online courses for teachers (Ministry of Education 2014). Science is also likely to gain a greater emphasis in the matriculation program for university entrants. As reported in the *Malaysia Education Blueprint Annual Report 2013* (Ministry of Education 2013b), matriculation program subject-matter experts together with Pearson/Edexcel have benchmarked the curriculum standards and assessment of Malaysia's matriculation program against the United Kingdom's A-Levels program. The five benchmarked subjects are biology, chemistry, mathematics, physics, and engineering studies (civil/electrical and electronics/mechanical). This development is seen as a step forward in enabling Malaysian matriculation candidates to gain entry into foreign university science programs.

References

Alwis, Caeser De. 2005. "Attitude of Form Two Students Toward Learning Science in English: A Case Study of Schools in Kota Samarahan." Paper presented at an educational research seminar, Batu Lintang Teachers College, September 15–16, 2005, http://www.ipgkbl.edu.my/portal/penyelidikan/seminarpapers/2005/caesarUITM.pdf.

Bevins, Stuart C., Mark Windale, Ong Eng Tek, and Bill Harrison. 2001. "Active Teaching and Learning Approaches in Science: Towards a Model for Malaysian Science Education." *Journal of Science and Mathematics Education in Southeast Asia* 24 (1): 12–28, http://www.recsam.edu.my/r&d_journals/year2001/2001vol24no1/12-28.pdf.

Idris, Noraini, Loh Sau Cheong, Norjoharuddeen Mohd. Nor, Ahmad Zabidi Abdul Razak, and Rahimi Md. Saad. 2007. "The Professional Preparation of Malaysian Teachers in the Implementation of Teaching and Learning of Mathematics and Science in English." *Eurasia Journal of Mathematics, Science & Technology Education* 3 (2): 101–10.

Lazim, Zanariah, Mohd Izar Kasturi Ibrahim, M. Al-Muz-zammil Yasin, and Haji Meor Ibrahim Kamaruddin. 2009. "Pedagogical Skills and Contents Mastery among Science Teacher Trainees at the Faculty of Education, Universiti Teknologi Malaysia," http://recsam.edu.my/cosmed/cosmed09/AbstractsFullPapers2009/Abstract/Science%20Parallel%20PDF/Full%20Paper/27.pdf.

Ling, Pick Yieng, and Rohaida Mohd Saat. 2013. "Use of Information Communications Technology (ICT) in Malaysian Science Teaching: A Microanalysis of TIMSS 2011." *Procedia—Social and Behavioral Sciences* 103: 1271–8. doi: 10.1016/j.sbspro.2013.10.456

Martin, Michael O., Ina V. S. Mullis, Pierre Foy, and Gabrielle M. Stanco. 2011. *TIMSS 2011 International Results in Science.* Chestnut Hill, MA: TIMSS & PIRLS International Study Center, Boston College, International Association for the Evaluation of Education Achievement (IEA), http://timss.bc.edu/timss2011/downloads/T11_IR_Science_FullBook.pdf.

Ministry of Education. 2006. *Integrated Curriculum for Secondary Schools.* Putrajaya: Author.

———. 2013a. "Vision and Aspirations." In *Malaysia Educational Blueprint 2013–2025 (Preschool to Post-Secondary Education).* Putrajaya: Author, http://planipolis.iiep.unesco.org/upload/Malaysia/Malaysia_Blueprint.pdf.

———. 2013b. *Malaysia Education Blueprint Annual Report 2013*. Putrajaya: Author, http://www.padu.edu.my/files/ar/PADU_AR_2013_ENG%20FULL.pdf.

Misbah, Hanizah, and Noordin Shaharom. 2008. "Tahap kefahaman kemahiran komunikasi dan mengeksperimen di kalangan pelajar tahun dua pendidikan Fizik merentas program pengajian" [The level of understanding of the communication and experiment-based skills of second-year physics education undergraduates across study courses]. Paper presented at a national seminar on science and mathematics education, Skudai, Johor, Malaysia, October 11–12, 2008 (in Malay).

Khalid, Mohamad Isa 2001. "Kemahiran proses Sains di kalangan guru pelatih Diploma Pendidikan Maktab Perguruan" [Practical science skills among teachers in teacher college diploma education: A preliminary study]. Education dissertation, Perlis Teacher Training College (in Malay).

Mokshein, Siti Eshah, Hussein Haji Ahmad, and Athena Vongalis-Macrow. 2009. *Towards Providing Quality Secondary Education: Training and Retaining Quality Teachers in Malaysia*. Bangkok: UNESCO, http://unesdoc.unesco.org/images /0018/001877/ 187765e.pdf.

Mukundan, Jayakaran, Vahid Nimehchisalem, and Reza Hajimohammadi, R. 2011. "How Do Malaysian School Teachers View Professional Development?" *Journal of International Education Research* 7 (2): 39–45.

Organisation for Economic Co-operation and Development (OECD). 2009. "Figure 1: Comparing Countries' and Economies' Performance." *PISA 2009 Results: Executive Summary* (OECD database). Paris: OECD Publishing, http://www.oecd.org /pisa/46643496.pdf.

———. 2012. "Snapshot of Performance in Mathematics, Reading and Science." *PISA 2012 Results in Focus* (OECD database). Paris, OECD Publishing, http://www.oecd .org/pisa/keyfindings/PISA-2012-results-snapshot-Volume-I-ENG.pdf.

Ramlan, Zarina Suriya. 2009. "Change in the Language of Instruction in the Teaching of Science in English." PhD diss., University of Malaya.

Salleh, Kahirol Mohamad, Nor Lisa Sulaiman, and Heidi Frederiksen. 2014. "Comparison of Teacher Licensing Between the United States of America and Malaysia: Implementation and Practical Implication." *Education Journal* 3 (3): 190–4. doi: 10.11648/j.edu.20140303.21

Subramaniam, Selvaranee. 2013. "Interrogation of Evidence-Based Learning in Science among Primary Students: A Case Study." Unpublished action research report, University of Malaya.

"World Bank: Worsening Education Obstacle to Malaysia's High-Income Hopes." *Malay Mail Online*, December 11, 2013, http://www.themalaymailonline.com /malaysia/article/world-bank-worsening-education-obstacle-to-malaysias-high -income-hopes.

Further Reading

Akcay, Hakan, and Robert E. Yager. 2010. "The Impact of a Science/Technology /Society Teaching Approach on Student Learning in Five Domains." *Journal of Science Education and Technology* 19 (6): 602–11. doi: 10.1007/s10956-010-9226-7

Almacen, Bayani Ursolino. n.d. "Teacher Training Program: The Malaysian Perspective." Academia.edu, https://www.academia.edu/428150/Teacher_Training _Program_The_Malaysian_Perspective.

Daniel, Esther and Noraini.Idris 2007. "Malaysian Science and Mathematics Education: Reflection and Reinvention." *Journal of Educational Issues* 30 (2): 65–83 (in Indonesian).

"EPP4: Expanding Private Teacher Training." Economic Transformation Programme, http://etp.pemandu.gov.my/Education-@-Education_-_EPP_4-;_Expanding _Private_Teacher_Training.aspx.

Fadzil, Hidaya Mohamad, and Rohaida Mohd Saat. 2014. "Enhancing STEM Education During School Transition: Bridging the Gap in Science Manipulative Skills." *Eurasia Journal of Mathematics, Science & Technology Education* 10 (3): 209–18. doi: 10.12973/eurasia.2014.1071a

Lee, Molly N. N. 2004. "Malaysian Teacher Education into the New Century." In *Reform of Teacher Education in Asia Pacific in the New Millennium: Trends and Challenges*, edited by Y. C. Cheng, King Wai Chow, and Magdalena Mo Ching Mok, 81–91. Dordrecht: Kluwer Academic Publishers.

Malakolunthu, Suseela. 2008. *Teacher Learning in Malaysia: Problems and Possibilities of Reform*, 2nd ed. Kuala Lumpur: University of Malay Press.

Meerah, Tamby Subahan Mohd. 1998. *Dampak penyelidikan pembelajaran sains terhadap perubahan kurikulum* [The impact of inquiry learning in the revised science curriculum]. Bangi: Penerbit Universiti Kebangsaan Malaysia (in Malay).

Mubarak, Amani. n.d. "Science Teacher Preparation: A Comparative Study of Malaysia and Singapore." Academia.edu, http://www.academia.edu/5349721/Science _Teacher_Preparation_A_comparative_study_of_Malaysia_and_Singapore.

Nor Aishah Buang, Lilia Halim, Tamby Subahan Mohd Meerah, and Kamisah Osman. 2007. "Development of Entrepreneurial Science Thinking for Malaysian Science and Technology Education." Paper presented at an educational seminar on engineering and the built environment, University of Kebangsaan, Malaysia (in Malay).

Nasri, Nufaradila, Mohamad, Zakiha Mohd Yusof, Shanti Ramasamy, and Lila Halim. 2010. "Uncovering Problems Faced by Science Teacher." *Procedia—Social and Behavioral Sciences* 9: 670–3. doi: 10.1016/j.sbspro.2010.12.215

Phang, Fatin Aliah. 2010. "Implikasi pemansuhan PPSMI terhadap pengajaran danpembelajaran di IPTA" [The implications of the repeal of PPSMI on teaching and learning in public higher education institutions]. *Buletin Persatuan Pendidikan Sains dan Matematik Johor* 19 (1): 75–83.

Syed Zin and Sharifah Maimunah. 2000. "Malaysia," http://www.ibe.unesco.org /curriculum/China/Pdf/IImalaysia.pdf.

Syed Zin, Sharifah Maimunah, and Keith M. Lewin. 1993. *Insights into Science Education: Planning and Policies in Malaysia*. Paris: UNESCO.

UNESCO. 2011. *Secondary Education Regional Information Base. Country Profile: Malaysia*. Bangkok: Author, http://unesdoc.unesco.org/images/0021/002147 /214736E .pdf.

13

Nauru

Penelope Serow and Neil Taylor

Teaching Science in Nauru[1]

Backdrop

The Republic of Nauru, formally known as Pleasant Island, is an island country located in Micronesia in the South Pacific. Nauru's population is approximately 10,000. For a country of only 21 square kilometers, it is known worldwide as one of the three great phosphate rock islands of the world. The mining of phosphate deposits provided substantial wealth to the island's inhabitants in the late 1960s and into the 1980s. For some of this period, Nauru had the highest per capita income of any sovereign state of the world. After exhaustion of the phosphate deposits in the 1980s, Nauru became well known in Australia when the Australian government opened a center there from 2001 to 2008 in order to assess asylum-seekers arriving by boat. The most recent center remains open and is known as an "off-shore processing center."

As a result of extensive mining, Nauru has very little capacity for industry. The large area that was mined is uninhabitable and requires the implementation of a massive rehabilitation program. Education remains a key ingredient in Nauru to enable the Nauruan youth of today to find innovative approaches to industry and development in their home country.

The school system in Nauru includes infant schools (play center and "prep," ages 3 to 5 years), lower primary school (Year 1 to Year 3, ages 5 to 8), upper primary school (Years 4 to Year 6, ages 8 to 12), junior secondary school (Year 7 to Year 9, ages 12 to 15), and a senior secondary school (Year 10 to Year 12, ages 15 to 18). In addition, there is one Catholic school that caters for prep to Year 7. In 2012/13, the Department of Education of the Republic of Nauru developed local curriculum documents for prep to Year 10 in the areas of language, mathematics, social studies, and science. Each document was developed through a partnership between external consultants and local counterparts.

The Department of Education also adopted the Queensland Certificate of Education (QCE) in 2013 as the qualification for the final two years of schooling in Nauru. These (Years 11 and 12) are post-compulsory. All students are able to proceed to Years 11 and 12, with the majority of students commencing upper secondary studies. However, the completion rate is low. Students who do not meet the QCE requirements at the end of Year 12 can continue to work toward their certificate. Their learning account remains open, regardless of their age, but credits do expire after nine years. All students who finish Year 12 receive a transcript of their learning account in the form of a "senior statement."

The QCE qualification is internationally recognized and includes senior school subjects and vocational education and training. To be awarded a QCE, a student needs to demonstrate a significant amount of learning, to a set standard and in a set pattern, while also meeting literacy and numeracy requirements. These requirements are measured in terms of credits. Credits are banked when the set standard has been met. Students must have at least 20 credits in the required pattern to be awarded a QCE.

Due to the relatively small population of Nauru and its physical remoteness, most students who want to enter post-school studies but remain in Nauru have to enroll in courses delivered by international providers. University studies of this nature often require the establishment of small onsite campuses and effort to deliver offshore versions of the institutions' courses. In attempts to keep the delivery manageable and affordable, some institutions present shortened versions of their courses in the form of intensive face-to-face sessions and modifications to written assessments. While these measures alleviate immediate problems, students may not experience the benefits of reflection and deeper learning of concepts possible when studying over a longer period of time.

School Science

The Nauru Science Syllabus (Government of the Republic of Nauru 2012a) is organized into two stages: prep to Year 6, and Year 7 to Year 10. Throughout the development of the syllabus, teachers and government officials contributed important ideas and feedback, which were then developed and coordinated by an external consultant. The approach underlying the syllabus focuses on what students are learning as opposed to the content to be delivered. The prep to Year 6 Nauru (Government of the Republic of Nauru 2012a, 3), aims to develop in students interest in science as a field of endeavor, curiosity about the physical and natural world, an understanding of basic science concepts, and skills in logical thinking and problem-solving as well as in science communication techniques and scientific inquiry. It also aims to provide students with knowledge about local and global environments and issues.

The six content strands of the prep to Year 10 Nauru science curriculum (Government of the Republic of Nauru 2012b) are life science, Earth science, environmental studies, energy and forces, matter, and time and space. A final strand titled "science and inquiry" runs across all of the strands. The Years 7 to 10 Nauru science syllabus has the same aims but with the addition of achieving the prerequisite knowledge and skills to undertake Years 11 and 12 science courses, thereby enabling a smooth transition for students into the QCE. The content strands in Years 7 to 10 are scientific method, life science, Earth and environment, energy and forces, matter, and time and space, thus providing avenues to explore physics, chemistry, and biology using the prep to Year 6 foundations.

Of particular interest is the impressive rate of uptake over the past two years in relation to students choosing science subjects as part of their QCE studies. Approximately 50 percent of senior students are selecting a science subject, with a particularly large number of students choosing chemistry and/ or physics. It is not unusual to have a class of 10 to 15 students in these subjects; a similar sized school in Queensland would struggle to find 6 students choosing chemistry and/or physics. The student and community interest in science is already strong and could be further enhanced by local role models in the scientific domain.

The Nauru Teacher Education Project

Rationale and Objectives

Developing Pacific Island countries (PICs) often battle with a lack of available local teachers. Statistics in 2008 identified an increasing gap in teaching capacity in Nauru wherein "only 9% of teachers have a degree, 6.4% a Diploma, 50% have a Certificate, and 34.4% have no qualifications" (Department of Education and Training, Republic of Nauru 2008, 21). Approaches to alleviate this problem in many PICs have included employing expatriate teachers to supplement teacher supply from local teacher education institutions. However, these institutions sometimes struggle due to remoteness and a lack of available resources, resulting in too few well-trained local teacher educators.

Currently, upper secondary science teachers in Nauru are most frequently drawn from Fiji and Australia. All three senior science specialists are expatriates. In addition, the Department of Education in Nauru has contracted a science advisor from Australia to support secondary teachers of science and maintain the smooth running of a new science laboratory at the junior secondary school. This approach has also helped teachers adopt an inquiry approach to the teaching of science in the secondary context.

The Nauru Teacher Education Project (NTEP) began in 2013 as a result of a need identified by the Nauru Department of Education to increase the number of qualified teachers in local schools and improve local teachers' pedagogical and content knowledge. Hence, the program targets two cohorts: existing (inservice) teachers and students who have not taught before (referred to as the preservice cohort). The NTEP was designed as a two-year Associate Degree in Teaching (Pacific Focus), with successful candidates having the option to undertake a further two years of study toward a Bachelor of Education (Pacific Focus). The courses are designed in accordance with the Australian Quality Framework and offer teaching pathways in early childhood education, primary education, and secondary education, with secondary education students further specializing in mathematics, science, English, social studies (history, geography, or business studies), or information technology (IT). Students, especially those in the inservice cohort, are awarded, where appropriate, prior credit for relevant units taken at other higher education institutions.

Technology and flexible delivery are key elements of the program, given the geographical isolation of the island, the limitations of the local communications infrastructure, and related logistical challenges. Before the program began, 20 state-of-the art laptop computers, a lockable laptop trolley, two printers, and associated IT hardware were freighted to Nauru, and Internet connection was established via five classroom modems in a dedicated program space provided at a local secondary school. The classroom and laptops are available to the students Monday through to Saturday, and each student is given an electronic storage device to store their individual work and assignments. Classroom sessions are timetabled to allow for unit-specific tutorials and support as well as independent study. All equipment purchased for delivery of the program will remain with the Nauru Department of Education for use beyond the specified project, a key element of the sustainability of the program going forward.

The online degree is enhanced through multi-modal delivery integrating Moodle platform participation with classroom tutorials, group collaboration, and unit-specific workshops delivered by visiting lecturers from the University of New England (UNE). This format is uniquely structured to maximize program flexibility, provide ample opportunities in support of multiple modes of learning, and allow for adaptation and continuity of learning despite interruptions to communications technology, which frequently occur on the island.

Extensive online and in-person support is essential to student learning and progress within the degree. Initial support was provided by an onsite project representative who helped establish the necessary infrastructure and assisted students through the completion of their enabling units and enrolment in the formal degree. At the commencement of the associate degree program, the level of onsite and online support was increased substantially with the addition of two full-time Pacific Education lecturers and two UNE-based support staff. These individuals liaise with UNE unit coordinators, provide additional support and feedback to the Nauruan cohort, and play an important role in contextualizing unit materials for the Nauruan context.

The continual presence of onsite lecturers enables a consistently high level of support for all units (which are undertaken online) and facilitates the development of a learning community among the cohort. While this level of support is integral to the implementation and early stages of the program, it would be costly to maintain long term. Because of this, students who successfully complete the program will be able to fill on-island support roles for future cohorts of teacher education students.

Candidates for the program were proposed by the Nauru Department of Education, and selections were further based on an English writing and comprehension test administered by UNE in Nauru. Admissions offers were awarded to 47 students, and an initial cohort of 41 students was admitted and undertook a four-week introduction in the form of an "academic culture" program to prepare them for life as a university student and the level of academic rigor expected in the course. Two instructors from UNE's Academic Language Centre administered the program. Prior to starting the associate degree in the first trimester of 2014, students enrolled in pathway-enabling units in mathematics and English to build the literacy and numeracy skills foundational to their study. The course rules of the associate degree also require students enrolling in secondary science education to take an additional enabling unit in science. A particular sustainability component is the program's mentoring aspect, which has the long-term aim of building a local team of teacher educators with the capacity to provide the continuous full-time on-island academic support needed to help later cohorts complete this Pacific-focused, Australian Quality Framework-compliant qualification.

During the delivery of the Associate Degree in Teaching (Pacific Focus), a constant priority is the capacity building of students in the program so that they can provide professional development opportunities to their professional peers and develop preservice teachers. Throughout the program, assessments involve preparation of presentations, workshops, community and network sharing of teaching ideas, and inservice teachers' contributions to in-school professional development sessions. Over time, the university presence in Nauru will, as planned, diminish, leaving the local teachers of Nauru to initiate and lead teacher professional development but still with the support of external educators, from a variety of contexts, when beneficial.

Professional Education and Training Issues

Significant cultural issues come into play for any initiative providing effective teacher education in the Pacific region. According to Thaman (2014), delivery of culturally appropriate teacher education requires the flexible provision of globalized content catering to both local contexts as well as international teaching cultures. This requirement involves a balanced reliance on cultural values through online and face-to-face deliveries, even to remote areas of the Pacific. Given the geographical nature of the Pacific (i.e., many small island countries dispersed over huge areas), recent advances in technology, the Internet in particular, should offer considerable logistical benefits to the region.

However, in an earlier paper, Thaman (1999) cautioned that using technological tools to deliver online teacher education programs can overvalue technological ways of learning, which in many cases are incompatible with the social and economic infrastructures of developing countries. Thaman pointed out that technology use is prone to interruptions in supply and communication, thereby placing significant limitations on the effectiveness of the learning that is taking place. This issue is exacerbated by students and teachers having to learn the new and different ways of communicating that IT allows. These limitations tend to disempower learners rather than develop autonomous professionals who can work independently yet easily form interdependent professional learning communities.

Another obstacle according to Hogan (2009) is that in Pacific locations where remoteness has made it difficult to develop the necessary quality of technology infrastructure, online learners must first master the use of such technology. Yusuf (2009) nevertheless believes that flexible delivery modes have the potential to overcome barriers caused by remoteness, natural disasters, lack of quality technology, and contextual alignment with each student's personal and academic needs. The Nauru program therefore attempts to undertake this compromise with what might be considered a "blended" or "hybrid" approach that takes account of both the cultural and technological needs of delivery.

Green and Reid (2004, 1) argue that "teacher education—like education research as well as schooling itself—should always be understood as situated practice." Situated learning and community of practice notions (Lave and Wenger 1991) are collaborative interactive teaching and learning approaches focused on addressing a particular issue of common interest over an extended period of time. Such notions seem to be an ideal foundation for teacher education, where educators come together to share ideas and experiences for addressing teaching dilemmas and reflecting on learning issues.

Mentoring has become a popular concept in education, particularly in the area of teacher education. The original meaning of the word "mentor" refers to a "father figure" who sponsors, guides, and develops a younger person (Ehrich, Hansford, and Tennent 2004). However, mentoring can take many forms, both formal and informal, and although generally viewed as positive, the process can lead to certain negative impacts, such as the mismatch between mentor and mentee in some instances. In Nauru, the onsite lecturers' mentoring role can be described as mentoring for academic study and mentoring into the teaching profession.

Effective and positive mentoring as outlined above incorporates a pervasive characteristic of Pacific culture, which Sanga (2005) describes as its relational aspect. Here, Sanga (2005) was describing the ideal format of foreign aid for the Pacific, namely developing equal relationships between the foreign and local cultures and constructing nurturing decision-making structures that enable the foreign aid to mentor and nurture the local Pacific counterparts. In the context of our present discussion, Sanga's notion encompasses the art of synthesizing the best of practice from both foreign and Pacific teacher education practices and processes that best fit the unique Pacific context. In teacher education, such a synthesis should result in a compatible globalized teaching culture that supports international as well as local curriculum content and delivery. This bilateral focus is important if future generations of Pacific countries are to function in a globalized society.

The global focus is also evident in the program's provision of a four-week professional experience session in Australia for each Nauruan student enrolled in the associate degree. This four-week experience enables a cultural exchange beneficial to both countries involved. It also lets participants become immersed in the international classroom and to reflect on the usefulness of the observed and practiced strategies and adaptations required.

Approximately 25 percent of the Nauruan cohort of students participating in the teacher education program are experienced teachers who have been teaching for five or more years. These teachers have chosen to upgrade their existing qualifications or to complete an initial teaching qualification at the associate degree level. Hence, 75 percent of the cohort are preservice teachers with minimal prior teaching experience. One of the main intended outcomes of the NTEP is to enable sustainability of professional development, which relies on local teachers in the program becoming academic and professional mentors. While professional development is currently provided by expatriate educators, local teachers should be in a position to provide professional development support during the final six months of their studies.

Of particular interest in relation to capacity building are the leadership roles that a few of the inservice teachers are demonstrating. Through the program, individuals have been able to demonstrate particular areas of strength that were previously unidentified or undeveloped. These strengths have included numeracy, creative arts, using technology as a teaching tool, lesson planning, and unit design. Other preservice teachers have demonstrated capabilities that indicate their suitability as area coordinators for specific key learning areas, such as language, mathematics, or science. Building these areas of strength in the primary context will provide a higher level of understanding among students preparing for the secondary years of their schooling.

During the teacher education program, the preservice science teachers enrolled in the program are given the opportunity to reflect on their journey as they develop as a teacher by responding to open-ended online questions specific to studying in Nauru. In addition, the students are interviewed individually. Both sets of responses indicate that the students are supporting one another and that many are working together as a learning community. The students often sit around the table together and participate in collaborative learning. This practice is very different to the experience of most online students in other contexts, who do not have at hand a system that facilitates collaboration.

Although course material is presented in English, it is quite common to observe the students explaining an idea or a task to one another in Nauruan. These explanations connect the students in the program who have specializations in key learning areas other than science and with varied levels of teaching and academic experience. In this regard, the students are beginning to mentor one another.

Due to the constraints of having students with varied responsibilities participating in face-to-face sessions and online collaboration at different times of the day, there are few opportunities for the more experienced teachers to pass on their expertise and experience about teaching and/or academic study. Data evaluating the program does provide evidence that the structure of the students' study timetable needs to be reorganized to better provide an environment where program participants can mentor one another on a regular basis by working together as a community of practice. There is a need to determine a means of forming a communication bridge between the inservice and preservice teachers in Nauru.

Responses provided by students six months into the two years of study do not provide evidence at this stage of the program fostering independence. The notion of mentoring will therefore be monitored closely over the next 12 months in order to ascertain changes in the levels of independence of the individuals enrolled as online students in the Associate Degree of Teaching (Pacific Focus) in Nauru.

Note

1. This chapter describes an intergovernmental project to develop endogenous upper secondary science teaching capacity in a Pacific microstate where science education at that level has long been reliant on expatriate personnel. It accordingly does not follow the same sequence that the other chapters in this book abide by. ~*Editors*

References

Department of Education and Training, Republic of Nauru. 2008. *Footpath 11 Education and Training Strategic Plan 2008–2013.* Yaren: Republic of Nauru Press.

Ehrich, Lisa C., Brian Hansford, and Lee Tennent. 2004. "Formal Mentoring Programs in Education and Other Professions: A Review of the Literature." *Educational Administration Quarterly* 40 (4): 518–40. doi: 10.1177/00131 61X04267118

Government of the Republic of Nauru. 2012a. *Nauru Science Syllabus: Prep–Year 6.* Department of Education, Republic of Nauru.

———. 2012b. *Nauru Science Syllabus: Year 7–10.* Department of Education, Republic of Nauru.

Green, Bill, and Jo-anne Reid. 2004. "Teacher Education for Rural-Regional Sustainability: Changing Agendas, Challenging Futures, Chasing Chimeras?" *Asia-Pacific Journal of Teacher Education* 32 (3): 255–73. doi: 10.1080/ 1359866042000295415

Hogan, Robert. 2009. "Attitudes of Indigenous Peoples toward Distance Learning in the South Pacific: An Empirical Study." Paper presented at the World Conference on Educational Multimedia, Hypermedia and Telecommunications, Honolulu, Hawaii.

Lave, Jean, and Etienne Wenger. 1991. *Situated Learning: Legitimate Peripheral Participation.* Cambridge: Cambridge University Press.

Sanga, Kabini. 2005. "The Nature and Impact of Educational Aid in Pacific Countries." In *International Aid Impacts on Pacific Education*, edited by Kabini Sanga and Ana Taufe'ulungaki, 17–46. Wellington: He Parekereke, Institute for Research and Development in Māori and Pacific Education.

Thaman, Konai Helu. 1999. "The Forgotten Context: Culture and Teacher Education in Oceania." *Directions: Journal of Educational Studies* 21 (1): 13–30.

——. 2014. "Towards Cultural Democracy in University Teaching and Research with Special Reference to the Pacific Island Region." In *Academic Migration, Discipline Knowledge and Pedagogical Practice*, edited by Colina Mason and Felicity Rawlings-Sanaei, 53–62. Singapore: Springer.

Yusuf, Javed, 2009. "Flexible Delivery Issues: The Case of the University of the South Pacific." *International Journal of Instructional Technology and Distance Education* 6 (6): 65–71.

Further Reading

Agostinho, Shirley. 2005. "Naturalistic Inquiry in e-Learning Research." *International Journal of Qualitative Methods* 4 (1): 13–26.

Awayaa, Allen, Hunter McEwan, Deborah Heyler, Sandra Linsky, Donna Lum, and Pamela Wakukawa. 2003. "Mentoring as a Journey." *Teaching and Teacher Education* 19 (1): 45–56.

Creswell, John W. 2009. *Research Design: Qualitative, Quantitative and Mixed Methods Approaches*, 3rd ed. Thousand Oaks, CA: Sage Publications.

Hargreaves, Andy, and Michael Fullan. 2000. "Mentoring in the New Millennium." *Theory into Practice* 39 (1): 50–6.

Heirdsfield, Ann M., Sue Walker, Kerryann Walsh, and Lynn A. Wilss. 2008. "Peer Mentoring for First Year Teacher Education Students: The Mentors' Experience." *Mentoring and Tutoring: Partnership in Learning* 16 (2): 109–24. doi: 10.1080/13611260801916135

Serow, Penelope, and Rosemary Callingham. 2011. "Levels of Use of Interactive Whiteboard Technology in the Primary Mathematics Classroom." *Technology, Pedagogy and Education* 20 (2): 161–73. doi: 10.1080/1475939X.2011.588418

Sharma, Akhila Nand. 1996. "A Reflection on Qualitative Research Methodology: A Fiji Experience." *Directions: Journal of Educational Studies* 18 (2): 31–44.

Shenton, Andrew K. 2004. "Strategies for Ensuring Trustworthiness in Qualitative Research Projects." *Education for Information* 22 (2): 63–75.

14

Fiji

*Salanieta Bakalevu, Mesake Dakuidreketi,
and Temalesi Maiwaikatakata*

Teaching Science in Fiji

Backdrop

Fiji is an archipelago of over 300 islands of varying sizes, of which 100 are inhabited. Two main islands, Viti Levu, with an area of 10,389 square kilometers, and Vanua Levu in the north, with an area of 5,534 square kilometers, occupy over 80 percent of the land area. Both the capital city Suva and Nadi International Airport are on Viti Levu.

Fiji has had a colorful history and development. It has come a long way since it was first ceded to Britain in 1874, the colonial government introduced Indian workers to work in the cane fields, and the country gained independence in 1970. The British colonial legacy and the influx of the Indian populace have had significant consequences on the nation's political history and the Fiji education system. At the end of 1995, the population of Fiji was estimated at 750,000, and the numbers of indigenous Fijians and Indians as approximately equal. Europeans, part-Europeans, Chinese, and other Pacific Islanders made up the rest. The population of the Fiji Islands in 2013 numbered just under 0.9 million, 70 percent of whom were living on Viti Levu.

Missionaries established the first schools in the early 1800s. The Methodist Mission established schools in the rural villages and taught literacy and numeracy in the local Fijian language. The Roman Catholic mission established centrally located schools managed by the Marist missionaries. Their first secondary schools catered mostly for Europeans, were academic in nature, and were staffed by European teachers. For the indigenous Fijians, post-primary education was mainly vocational in nature and focused on the training of nurses, primary school teachers, and church ministers. European teachers taught in English, and the English language quickly became the medium of instruction. The Indo-Fijian religious organizations established their own schools in the 1890s. Today, about 18 different such organizations, including religious groups, administer some of the best schools and post-secondary institutions in the country.

The pattern of ethnic difference in academic performance is a long-standing one in Fiji. Various studies (Dakuidreketi 2014, 2004; Otsuka 2006) have identified issues associated with culture and belief, demography, economic status, and resources as major contributing factors. To reduce this gap and improve the quality of Fijian education, Fiji's government deemed it necessary in later years to introduce affirmative action policies in favor of the indigenous Fijian and Rotuman groups (the natives of the island of Rotuma, which is under the jurisdiction of the Fiji government). In a study of the effects of the affirmative action, Qovu (2013) reported that the Ministry of Fijian Affairs Scholarship Scheme and investment in selected government secondary schools, called centers of excellence, had yielded positive outcomes in terms of learning facilities and academic achievement. His study revealed that "for the Fiji Seventh Form Examination (Year 13) alone, the pass rate of indigenous Fijian students had improved from 63.5 percent in 2005 to 73.5 percent in 2006" (Qovu 2013, 236).

The military coup of 2006 saw a military-backed regime lead the country until national elections in 2014. The regime established an education policy in 2013, supported by the 2013 national budget that stipulated the provision of "free education for primary and secondary students as well as loan schemes for tertiary students at extremely low interest rates" (Fiji 2013, 3). Additional items covered under this policy included the expansion of basic compulsory education to 12 years of schooling, the removal of Years 6, 8, and 10 external examinations, the introduction of competency-based assessment, the provision of free textbooks to all primary schools and progressively to secondary schools, the provision of transport assistance to students, and a zoning policy for school intakes and enrolments to bring education closer to children's homes.

Other important policies were continuing government assistance for the upgrade of facilities in non-government schools, mainstreaming vocational courses through the introduction of the Basic Employment Skills training program, expanding and upgrading rural high schools up to Form 7 (Year 13) so that rural students could more easily access university and advanced technical education, introducing e-learning, and reviewing the curriculum through the formulation of the Fiji National Curriculum Framework (NCF). The review of the NCF in 2013 chartered a new direction for the provision of education to meet the needs of all children and the national workforce. The framework is now organized under six foundation areas of development (FALD) for early childhood education and seven key learning areas (KLAs) for the primary and secondary levels. English, mathematics, and science are important KLAs.

National elections on 17 September 2014 saw Fiji return to democratic rule since 2006 and the party of Commodore Voreqe Bainimarama winning a comfortable majority. While most of the items listed under the 2013 education policy remain intact, one significant change will be the reintroduction in 2016 of the Year 6 and Year 10 external examinations. Concerns raised by stakeholders including parents and industries about the quality of students coming out of high schools has prompted this move.

Education in Fiji is a partnership between government and communities. Fewer than 2 percent of schools in Fiji are government schools. Instead, the majority of schools are controlled by communities and church organizations, while the rest are privately run. Of the 178 secondary schools in Fiji, only 12 are government owned. All non-government schools are managed by properly constituted controlling authorities. While this unique partnership in the management of schools promotes a sense of identity and common purpose, it also has a bearing on the critical choices that schools make, such as those relating to subject options and combinations.

Primary/elementary education runs from kindergarten to Year 8, followed by secondary education from Year 9 to Year 13. There are two possible exit points for secondary education: at the end of Year 12 with the Fiji School Leaving Certificate (FSLC) or at the end of Year 13 after the Fiji Seventh Form Examination (FSFE). Examinations at all levels serve rigorous selection purposes; those students who succeed continue to the next level while those who fail have the opportunity to repeat the year. Of the 178 secondary schools in Fiji, about 90 percent offer Years 12 and 13. There is now an equal distribution of these schools in the urban areas as well as the rural and remote areas. This distribution is an important consideration in terms of availability of science laboratories and science equipment, especially for the specialized sciences.

Enrollment data over the years have shown that only about 50 percent of Year 12 students enroll in Year 13 because of the large dropout rate after the Year 12 examinations. The expansion and upgrading of rural high schools up to Form 7 (Year 13) was an attempt on the part of the government to curb the rural–urban drift and provide easier access for rural students to university and advanced technical education. A study of enrolment trends up to 2013 indicates a larger number of male than female students enrolling in secondary education. However, the trend changes in Form 7, where females make up about 58 percent of the roll. Over the years from 2007 to 2013, the pass rate in the FSFE averaged around 75 to 80 percent of the cohort. Of these students, about 30 percent qualified for university study, having attained the minimum cut-off marks of 250 out of 400.

Fiji hosts three universities: the University of the South Pacific (USP), the Fiji National University (FNU), and the University of Fiji (UOF). There is also a technical-vocational post-school sector.

School Science

At the primary level, all students study elementary science from Years 1 to 6 and then take basic science in Years 7 and 8. Students continue to study basic science up to lower secondary level (Years 9 to 10). From Year 11 to Year 13, students are required to take five subjects that they can pick from a combination of English and mathematics as the compulsory subjects and three others chosen from biology, chemistry, and physics as pure science options, and technology subject options (technical drawing, home economics, agriculture science, introduction to technology, and computer education). Students may choose combinations of science and technology subjects to fulfil their five subject requirement at Years 11 to 13. Other subject options for social science and commerce are history, geography, accounting, and economics. The combination of subjects is usually determined by the individual schools, with that determination depending mainly on the availability of qualified staff for the subject, students' prior academic achievement, class sizes, and school location.

Science is one of the seven KLAs at primary and secondary levels. Fiji's Curriculum Advisory Service has been progressively revising science curricula by translating the old "prescriptions" into syllabuses. The new syllabuses clearly identify KLA outcomes, strand and sub-strand outcomes, content learning outcomes, and expected student achievement indicators. The new syllabus for Year 11 physics was trialed nationally in 2014, and the Year 12 physics syllabus in 2015. Implementation of the new chemistry and biology syllabuses for Years 11 and 12 began in 2015. However, the prescriptions for the three science subjects for Year 13 remained unrevised.

In terms of student enrolments into different subjects at the Year 13 level, the Ministry of Education's 2013 annual report (Ministry of Education 2011) showed that more females than males were enrolled in chemistry (1,800 females and 1,000 males) and biology (1,900 females and 700 males), while the reverse was recorded for physics (700 females and 1,000 males). There is greater concern about the relatively low numbers of all students opting for physics. This does not augur well for the need to improve the number of physics teachers. Similar trends were observed in the pass rates of the Year 13 national examination. In 2011, the results reported in the annual report for that year (Ministry of Education 2011) indicated that some 90 percent of candidates sat the mathematics paper, 41 percent the biology paper, 36 percent the chemistry paper, and only 25 percent the physics paper. The proportion of students who scored between Grade 1 (excellent) to Grade 5 (satisfactory) and were expected to progress to university studies were 72 percent for mathematics, 65 percent for biology, 67 percent for chemistry, and 70 percent for physics.

Teacher Academic and Professional Education and Training

The University of the South Pacific is the leading provider of secondary teachers in the region. USP offers four-year concurrent preservice BSc/Graduate Certificate of Education (GCED), BA/GCED, and BCom/GCED programs for secondary teachers, and an in-service Bachelor of Education (BEd) program in early childhood, primary, and secondary education. The Fiji National University (FNU) offers two preservice programs for secondary teachers—a Higher Diploma in Education (Secondary) and a BEd (Secondary).

The programs at both institutions require secondary teacher trainees to study two subject majors. In science, biology and chemistry are the most popular combination; the physics and mathematics, physics and computer science, and physics and information systems combinations are less popular. The four-year preservice teacher education program prepares specialist teachers for teaching two specialized subjects of their choice in secondary schools throughout the South Pacific region. The combined BSc/GCED, BA/GCED, and BCom/GCED programs require students to have successfully completed 7 to 8 courses in discipline majors, 11 education courses (including a 17-week practicum experience), 4 additional university courses, and enough electives to make up 31 courses before they graduate.

Compared to teachers in other Pacific island countries, teachers in Fiji are fairly well qualified. In a report on teachers and teacher education in Fiji, Tagivakatini (2007, 39) reported an "almost 100% teacher-trained teaching force." The teacher classification by qualification in Fiji indicates that in 2013 there were 4,650 teachers in secondary schools and that the majority of these teachers had appropriate qualifications. The Ministry of Education's 2013 annual report recorded that close to 60 percent of secondary teachers had at least a Bachelor's degree qualification while 38 percent had a diploma in teaching and about 2 percent had a certificate-level qualification.

Continuing Professional Development

Most science teachers in Fiji have a BSc degree with a teaching qualification (either USP's BSc /GCED and BEd or FNU's BEd). A small minority have only a BSc degree without any teaching qualification. The members of this latter group usually return as inservice trainees to one of the institutions to acquire the GCED qualification. Some choose to study part time and others full time.

Teacher education institutions in Fiji continue to provide professional development to teachers through their inservice programs. In the past, these institutions also provided additional services in answer to requests from individual schools and sections of the Ministry of Education. These activities ceased when, in 2013, the Teacher Registration Board required all teachers and other educational professionals and teacher educators who wished to carry out activities in schools to be registered annually. The administrative requirement for annual registration is a tedious exercise that includes securing medical and police clearance. At the same time, the Ministry of Education set up its own Professional Development Services Unit to support schools with their continuing professional development activities. The major losers in this exercise were specialist groups such as the science and mathematics teachers who relied heavily on specialist educators in the universities.

Science Teacher Professional Associations

Although Fiji does not have a science teachers association, some secondary school teachers are members of local or regional associations, such as the Fiji Mathematics Association and the Chemical Society of the South Pacific.

Issues in Upper Secondary Science Teacher Quality

Teacher Supply

Figures provided in the Ministry of Education's annual reports indicate acceptable teacher–student ratios of around 1:30. However, reports from schools suggest that ratios of at least 1:40 are still common. Inservice trainee teachers continually talk about the challenges of teaching large classes in small classrooms, a situation that leaves little room for innovation. Some reports warn that the ministry will be forced to raise the teacher–student ratio to as high as 1:45 if additional teachers are not found.

Traditionally in Fiji, teaching has been seen as a stable profession that one could stay in for life. While this may still be true, there is no denying the increasing number of teachers who continue to leave the profession for various reasons. The heavy workload of teachers is an important factor. Teachers in Fiji carry an average teaching load of 27 to 30 hours per week. For science teachers, teaching does not offer the kind of incentives that are available in science-based occupations elsewhere. A salary package that is not competitive, that presents limited opportunities for advancement, and comes with a demanding workload and a stressful working environment are major challenges. Many science teachers are themselves technicians and laboratory managers all in one. The lack of necessary equipment and resources is a source of frustration and huge disappointment for science teachers. Over the years, many science graduates have left the teaching profession after the first or second year to join private-sector companies such as the Fiji Sugar Corporation, Flour Mills of Fiji, or the mining companies.

A large number of science teachers resigned and emigrated after the 1987 military coup. A 2008 analysis of the education sector identified high teacher attrition as one of the major concerns. A 2009 directive from the government to lower the retirement age of civil servants from 60 years to 55 years added salt to the wound. This unexpected turn of events affected the teaching profession in a big way because many heads of department and senior experienced teachers were suddenly retired without a succession plan in place. Fortunately, that policy has been relaxed slightly to allow for proper handing over and a smoother transition between the old and the new so as not to adversely affect teaching and the organization of schools.

Professional Education and Training Issues

The preservice teacher education program culminates in formal graduation and the posting to schools. A related part of this process is the induction of new teachers, something that has not been given due consideration. We believe that teacher education providers such as USP and FNU should work with the Ministry of Education to induct teachers properly into the profession, a suggestion that the Ministry of Education is actively considering. The induction process will serve to properly orient teachers toward the Fijian educational context and adequately equip them with a sound knowledge of the ministry's ideals and policies.

The shortage of physics graduate teachers is something teacher education providers also need to address. In 2014, 47 preservice teachers who graduated with a BSc/GCED were chemistry and biology majors; only 15 had physics as one of their majors. Of the 19 who completed the GCED qualification in 2013, 10 were science major students, all 10 of whom had either a chemistry or biology major, meaning of course that none of them had a physics major.

Issues at the Chalkface

Science teaching at higher levels is changing very slowly. We (the authors of this chapter) interviewed a group of inservice trainee science teachers enrolled in USP's GCED program and asked them to tell us how they were teaching science and what methods they commonly used in their science classes. A large number said that the traditional form of teaching with emphasis on learning scientific knowledge from textbooks is still the dominant approach in school. They advised us they were teaching from the Curriculum Advisory Service prescriptions and were always mindful of the required amount of work to be covered for the national examinations. (A specified proportion of the syllabus as well as a defined number of practical experiments have to be covered by a particular date.)

It was encouraging, however, to hear the teacher trainees talk about their efforts to vary teaching methods and make science classes interesting and enjoyable for students. Their comments indicated that small-group discussions are increasingly being used and are improving students' confidence and capacity to learn and make sense of science as they enter into dialogue with peers and the teachers. Biology teachers talked about class discussions, question-and-answer sessions, and students attempting to make their own summary notes. Chemistry teachers said that while they continued to use teacher talk and note taking, they were also using whole-class discussions and question-and-answer sessions.

Teaching resources are mostly a function of how much funding each school devotes to them. Considering that most schools are non-government, the level of resources expended to support teachers' work varies greatly from school to school and from management to management. Fundraising is a common feature in Fiji schools, and funds raised this way are used for operational costs and other school expenses or projects.

All schools that offer science at the upper secondary school level have a laboratory. However, the non-availability of even the most basic equipment for use in demonstration and laboratory exercises continues to be a limiting factor. Laboratories in the urban schools are better equipped and in better condition than those in rural settings. A science teacher in an island school mentioned that the laboratory in the school did not have storage space to keep equipment, some of which had been donated. The same school does not have basic equipment, such as the delivery tubes necessary for most tasks that involve gas production, as well as chemicals such as salicylic acid and pentan-1-ol for the preparation of organic compounds. Then there are other remote schools where the supply of electricity is unreliable and intermittent. In these schools, it goes without saying that some of the practical elements of curricula that require the use of microscopes with high resolution are a challenge. While teachers have been trained to improvise because they serve in difficult environments, there is no denying that challenges do affect the quality of learning and teaching.

Practical work is an integral part of learning science. The majority of the preservice teachers surveyed indicated that in their previous schools, practical work was conducted once a week for each science subject; a smaller proportion indicated that practical work was hardly ever conducted because of a serious lack of apparatus and resources. A student teacher who came from a committee school reported that more often than not their teacher wrote everything on the board and they copied notes. In such circumstances, teachers discussed what should have been done, what was expected, and what the conclusions should have been. Of course, these circumstances deprive students of the opportunity to actually carry out the experiments, which would give them a sense of ownership and confidence and therefore positive feelings about science and learning.

A survey of new students to USP that we conducted inquired about common practices in laboratory classes in their previous secondary schools. A large number reported that it was uncommon for experiments to be done by individual students. Most students reported that, with respect to chemistry and physics, they never participated at all in the laboratory practical sessions. Most times, the experiments were done in groups or demonstrated by the teacher. The students who experienced group work said that doing experiments in groups did not really develop learning because not everyone was involved. They also said it was not uncommon for some students to lose interest and withdraw completely from the activities. Some students said they preferred to observe only because they were unsure about using the equipment. A lack of confidence and fear of experiments and mishandling science equipment was seen as a common problem arising out of the lack of practice and training.

These discussions with teacher trainees indicated that, despite the challenges, they all believed in the immense value of practical work, seeing it as the most useful aspect of science learning because it helps students under-stand science and its concepts. It is important that science teachers make room for scientific inquiry as a central part of science instruction. They can do this by decreasing their emphasis on lectures and didactic teaching. Much teacher practice during laboratory classes is dictated by the type of inquiry promoted in curriculum documents. In Years 11–13, structured inquiry is still prevalent. The current laboratory manuals for all science subjects that have been developed by the Curriculum Advisory Service allow for students to develop manipulative skills; collect, communicate, and interpret data; and formulate conclusions based on data. An additional requirement is for students to have the opportunity to formulate questions, identify and control variables, and plan their own investigations—all integral skills for scientific inquiry. This approach is likely to be strengthened when the laboratory manuals are revised in line with the new syllabuses, all of which emphasize student engagement in open scientific inquiry.

A major limitation that is noted particularly in the rural and remote schools is the availability of qualified staff for the science classes. A difference in teacher quality, as measured by qualification and experience, exists between rural and urban areas. In those areas where there are no qualified science teachers, the next-best option has been to hire teachers who have a Diploma in Education (Science) qualification or those who have a straight BSc degree. It is also not uncommon to use teachers who are university trained though not in the required subject area to fill the gap. When the aim is quality teaching, all three groups of teachers present challenges: where one group does not have the required level and depth of science knowledge, the other does not know how to teach, let alone teach the subject. Such teachers are likely to teach in the way they were taught. In these cases, didactic teaching from a textbook and the science manual is the norm, with students required to copy notes from the blackboard and textbook and little opportunity to practice scientific inquiry. The limited opportunity for scientific inquiry for both teachers and students is a serious concern.

Language issues have always been problematic for Fiji students, for whom English is either a second or third language depending on the dialect of the area. The missionaries introduced English in the curriculum in the early 1890s. As the colonial government took over education, the use of local vernaculars as languages of instruction decreased (Mugler 1996). The language of instruction in all secondary schools is English: students are taught in English, use English textbooks, and write examinations in English. Students are not allowed to speak in their native tongue while at school except during vernacular and cultural activities.

We are aware that the local vernacular languages lack the words to carry the meaning of scientific and technical concepts. While most Fijian teachers feel uncomfortable and inadequate teaching in the English language and would prefer to do so in their mother tongue, the challenge is more for science and mathematics teachers, who need to find alternative ways to put across the technical concepts of their subjects. Many students, especially those in the rural and remote areas, express the same sentiments. The reality then is that local teachers and students are communicating in a language they are not proficient in and are learning technical scientific language that is mostly unrelated to their mother tongue. Science students therefore have to learn both the language of instruction and the language of science. Their problems are aggravated when dealing with concepts such as work, force, pressure, energy, and animals, each of which has a different meaning in everyday life and in the science classroom.

Studies carried out by local researchers (Dakuidreketi 2004; Muralidhar 1991) suggest that one of the reasons both primary and secondary science students have not been performing well in science is difficulty with science terminology, which forces them to memorize definitions and other facts that have no relation to their prior knowledge or meaning in their mother tongues. One of the major outcomes of this dilemma is one-way instruction in schools—the teacher teaches and students listen and copy notes. Students are shy and afraid to ask questions when in doubt and do not respond to questions even when they may know the answers. The challenge for teachers is therefore one of creating a learning environment where students' ideas, in whatever form, are listened to, and where they are encouraged to engage in dialogue with one another and their teachers.

How do we address this issue? Some proposals include bilingual science education and approaching science as a language. There is no denying that the use of more than one language to express scientific ideas adds value in itself. The development of a Fijian glossary of scientific terms as well as a dictionary or text of science terms in Fijian is a project already discussed by local science educators.

The use of ICT in teaching is an important development that the Ministry of Education and the government have been working on for at least two decades. Since 1998, the Ministry of Education has actively pursued integration of ICT into teaching and learning and the development of students' digital literacy, primarily through its ICT in Schools program. The program encompasses the provision of essential ICT infrastructure in schools, including access to broadband connectivity, continuous professional development for teachers in ICT, integrating ICT within the curriculum, and providing curriculum-relevant digital content and software. Computer science is one of the subject options for the FSLC. Initially, the subject focused mainly on hardware and the use and care of computers. However, the course has since been revised to include the effective use of technologies in teaching and everyday life. Computer science is now also taught and examined at Year 13.

In 2011, the Fijian government introduced the use of Internet and extranet into its 12 secondary schools, although many non-government schools had made this move earlier and were already using e-services. In the same year, the ministry assisted secondary schools to develop school websites that link into the secondary website at the central government administration. By 2013, a total of 91 secondary schools had developed school websites linked to this central site. This development was a critical one because it opened up access of communication, facilitated more dialogue, and improved decision making processes and implementation of activities.

More recently, the government has established several telecenters in townships and critical locations close to rural communities to open up telecommunications and travel services to Fiji's rural and remote areas. These have enabled the hoped-for easier means of communication, including telephone and Internet services. In addition, some non-government schools have received much-needed computers and additional e-resources to boost their ICT facilities. Some communities have set up their own telecommunication infrastructure to provide for their telecommunication services.

Continuing Professional Development Issues

While teacher education programs continually introduce teacher trainees to new philosophies and practices, graduate teachers go into teaching environments in schools and confront challenges to do with distance and access, leadership and management, school ethos, culture of teaching, and limited facilities that do not allow for innovation and change and which, in effect, push them to "fit in" with the status quo. It is for this reason that the new Minister of Education's plan to have Fiji's teacher education providers partner with Ministry of Education officials in ongoing professional development of teachers is progressive. This notion of shared responsibility holds much promise.

The increasing number of teachers, including secondary science teachers, who are upgrading their qualifications to a Master's degree is also very encouraging. However, many are doing this for migration purposes or for more lucrative opportunities outside of teaching. The salary structure shows that teachers receive a salary increase after the first degree, but no other incentive is provided for further qualifications after that.

Trends and Developments in Upper Secondary Science Teacher Quality

The provision and regulation of teacher education in Fiji has undergone major changes. Initially, the Ministry of Education had considerable input into regulation of teacher education, including the selection of intakes as well as the nature of the programs. With decentralization and privatization of government services, programs are now at the discretion of the providers, and it is they who make the decisions about organization and quality. However, a National Qualifications Authority is now also in place and is working to establish benchmarks and standards to be met by all providers. The authority will thus serve as a quality assurance mechanism for monitoring the quality of teacher education programs. Developing a science teacher standard will be the next step forward in the attempt to improve the quality of science teaching in secondary schools.

If the words of the new Minister of Education are anything to go by, the professional development of teachers will become not only more structured and continuous but also the shared, collaborative responsibility of both the teacher education providers and the Ministry of Education. In effect, this aspect of professional development is not really a new idea, as teacher education institutions provided these services in the past. In the meantime, Ministry officials are conducting presentation sessions to inform the teacher education providers of the changes and also implementation of Fiji's new National Curriculum Framework.

References

Dakuidreketi, Mesake Rawaikela. 2004. *Contexts of Science Teaching and Learning in Fiji Primary Schools: A Comparative Study of Ethnic Fijian and Indo-Fijian Communities.* PhD thesis, University of Canterbury, Christchurch, New Zealand.

Dakuidreketi, Mesake Rawaikela. 2014. "Scientific Method and Advent of Literacy: Towards Understanding of Itaukei and Indo-Fijian School Students' Differential Achievement in Science." *Universal Journal of Educational Research* 2 (2): 99–109. doi: 10.13189/ujer.2014.020201

Ministry of Education. 2011. *Annual Report.* Suva: Author, http://www.education.gov .fj/images/AnnualReports/Annual%20Report%202011.pdf.

Ministry of Education, National Heritage, Culture and Arts Annual Report. 2013. Suva: Author, http://www.parliament.gov.fj/getattachment/Parliament-Business /Annual-Reports/Education_2013.pdf.aspx.

Mugler, France. 1996. "Vernacular Language Teaching in Fiji." In *Pacific Languages in Education,* edited by France Mugler and John Lynch, 273–87. Suva: Institute of Pacific Studies.

Muralidhar, Srinivasiah. 1991. "The Role of Language in Science Education: Some Reflections from Fiji." *Research in Science Education* 21: 253–262.

Otsuka, Setsuo. 2006. *Cultural Influences on Academic Achievement in Fiji.* Paper presented at the AARE Annual Conference, Melbourne, Australia.

Qovu, Emasi. 2013. *A Critical Review of the Affirmative Action Policy in Education of the Indigenous Fijians 1987-2006: Policy, Rationale and Implications.* PhD thesis, The University of the South Pacific, Fiji.

Tagivakatini, Sereana. 2007. "Teachers and Teacher Education in Fiji." In *Teachers and Education in the Pacific,* edited by Seu'ula Fua and Kabini Sanga, 34–52. Suva: Institute of Education, University of the South Pacific.

Further Reading

Aalbersberg, William G. L., Muralidhar Shrinivasiah, and U. Singh. 1996. "Chemical Education in Fiji." In *Chemical Education in Asia-Pacific,* 67–93. Kuala Lumpur: The Chemical Society of Japan and the Federation of Asian Chemical Societies.

Giddings, Geoffrey J., and Bruce G. Waldrip. 1993. *Teaching Practices, Science Laboratory Learning Environment and Attitudes in South Pacific Secondary Schools.* (ERIC Document No 363 507), http://files.eric.ed.gov/fulltext/ED363507.pdf.

Mangubhai, Francis. 1982. Introduction. In *Duivosavosa: Fijian Languages, Their Use and Their Future,* edited by George Bertram Milner, David G. Arms, and Paul Geraghty. Museum Bulletin No 8, 1–2. Suva: Fiji Museum.

Ministry of Education, National Heritage, Culture and Arts Annual Report. 2012. Suva: Author, http://www.education.gov.fj/images/AnnualReports/ Annual%20 Report%202012.pdf.

Pennington, Bill, Nelson Ireland, and Wadan Narsey. 2010. *Fiji Education Sector Program: Independent Completion Report* (AidWorks Number INF528). AusAID, http://www.oecd. org/countries/fiji/48473721.pdf.

Prasad, Surendra Bindesri. 1996. "Teaching Form 7 Physics in Secondary Schools in Fiji: Current Difficulties and Some Proposed Solutions." *Directions: Journal of Educational Studies* 18 (2): 67–75.

Voigt-Graf, Carmen, Robyn Iredale, and Siew-Ean Khoo. 2009. *Teacher Mobility in the Pacific Region Cook Islands, Fiji and Vanuatu & Australia and New Zealand.* Australian Research Council Report, http://www.apmrn.usp.ac.fj/fileadmin/files /publications/Teacher_mobility_ARC_Report_20091.pdf.

15

Indonesia

Ari Widodo, Diana Rochintaniawati, and Riandi

Teaching Science in Indonesia

Backdrop

Indonesia is an archipelagic country of 17,504 islands. With a population of approximately 250 million, it is the fourth most populous country in the world. The country is home to more than 300 ethnic groups with different cultures and languages (Central Bureau of Statistics 2013). The national language is Bahasa Indonesia, but most people also speak their ethnic language mother tongues. As reflected in the national slogan, "Unity in Diversity," Indonesia is truly a country of variety.

The diversity of the country is reflected in the education system. Education is provided by the government and private sector. The government is represented by three ministries: the Ministry of Education for preschool up to senior high school, the Ministry of Research and Higher Education, and the Ministry of Religious Affairs. The private sector is represented by a variety of foundations. With the permission of the Ministry of Education, foundations can run schools and universities. Because foundations are founded by various community groups (e.g., religious, ethnic, community associations), schools may vary as a reflection of the foundations they are affiliated with.

An independent board called Badan Standar Nasional Pendidikan Indonesia (National Education Standards Agency of Indonesia) develops, monitors, and evaluates educational standards in Indonesia. The board has eight sets of standards in place for education: graduate competencies, program content, instructional processes, educators and supporting staff, facilities, management, funding, and evaluation (National Education Standards Agency of Indonesia 2009). All schools are required to adhere to these standards in their operations. The Ministry of Education's school accreditation board assesses compliance with these standards.

Kindergarten or preschool is not compulsory, but many children attend kindergarten. Indonesia has three levels of schooling: elementary school (Grades 1 to 6), junior secondary school (Grades 7 to 9), and senior high school (Grades 10 to 12). Compulsory education runs for nine years, which means that children are supposed to complete junior high school. Under the current curriculum, students have to choose one of four tracks on entering Grade 11. These are science, social studies, language, and religious education.

At the end of each school level, students have to sit national examinations in government-prescribed subjects. Science is considered an important subject and is included in the national examinations. Students' achievement in the national examinations is used as one of the criteria for graduation and for admission to government secondary schools. The transition rate from primary school to lower secondary school is 82 percent, while the transition rate to upper secondary school is 97 percent (Ministry of Education and Culture 2012). Admission to higher education is not based on scores on the national examination but on selection tests. Selection processes for entry to public universities are managed by a consortium of those universities, while selection processes to private universities are conducted by the institutions themselves.

School Science

Science as a school subject appears in the curriculum from Grade 4, before which science is meant to be integrated with other subjects. However, in many cases, schools teach the subject separately, as they do for the higher grades. At primary level, science is taught for four school periods of 35 minutes each week. One of the issues in science teaching at primary school is the level of teachers' competence in the discipline. In many cases, teachers do not have sufficient background knowledge for teaching science.

At junior secondary school, science is allocated four periods (of 40 minutes) each week. Although the subject is called science, it tends to be taught as chemistry, physics, and biology lessons rather than as integrated science. Initiatives directed toward teaching science as an integrated subject had little success. For one thing, teachers were not properly prepared to teach science as an integrated subject because teacher education at the universities in the past trained preservice teachers as biology, chemistry, or physics teachers. For another, the curriculum does not actually present science as an integrated subject, as the division of topics is along disciplinary lines.

At senior secondary level, science is presented as three different subjects (biology, physics, and chemistry). During the first year, each science subject is allocated two 45-minute periods. In the second and third years, science is taught only in the science track, at which time each subject is allocated 4 periods per week, that is, 12 periods of science per week. In most schools, science is the most popular track. The science track is generally perceived as the most prestigious track and only for "smart" students. A study on students' interest in science conducted by Suratno, Widodo, and Kadarohman (2013) showed that Indonesian students are highly interested in science. Among the reasons why students choose the science track at school is the opportunity to obtain a secure job and earn a good income. In the Indonesian system, students who take the science track at upper secondary level can also take non-science studies such as economics or law at university. Students in the non-science tracks can undertake only non-science studies.

Teacher Academic and Professional Education and Training

In the past, teacher education for primary school teachers was conducted separately from teacher education for high school teachers. Until the 1980s, teachers for high schools were prepared at the university level, while pre-service primary school teachers attended three-year colleges, equivalent to senior high school in the current system. Toward the end of that decade, the government set out to improve the quality of education, including the quality of teachers. The government mandated that all teachers should be university graduates, which led to the universities taking over the responsibility for primary teacher education. These universities included specialized education ones, such as the Indonesia University of Education, as well as universities operating faculties of education. As previously mentioned, communities may also run schools and universities, and these private universities, in turn, can also conduct teacher education.

The minimum education standard for teachers keeps changing. In the 1980s the minimum requirement to be a teacher was two years at university. At the end of the 1990s, effort was made to lift the requirement for teaching to four years of education at the university level. The enactment of the law on teachers and lecturers (Government of the Republic of Indonesia 2005) set an even higher standard for teachers by requiring teachers to have four years of education at university and take part in a certification training program made up of subject matter, pedagogy, and teaching practice components.

The new system provides two pathways to become a teacher. The first basically provides an additional training program for existing graduates of teacher education, while the second pathway is a completely new track that allows graduates of non-education majors to become teachers by undergoing professional training involving pedagogy courses, subject-matter pedagogy, and teaching practice (Indonesia University of Education 2010). The second track for becoming a teacher is not yet common because the traditional concurrent model of teacher education, in which content and pedagogy are taught in the same program, still dominates.

The multiplicity of teacher education providers (public and private, universities of education, and faculties of education within universities) means that quality assurance is a difficult issue. Although there is an accreditation board that plays a role in quality assurance control, the variation in quality amongst these providers is a serious issue. Accreditation of an education program does not necessarily reflect the quality of the graduates, however. The three levels of accreditation status—A (very good), B (good), and C (fair)—reflect the significant variability in graduate quality across the nation.

Continuing Professional Development

Inservice teachers can access professional development services through a number of sources. The government plays the main role in this regard through agencies such as the National Science Teacher Training Center and the programs managed by local governments. Professional development services are also available through community agencies, teacher associations, and universities. Many universities provide training for teachers as part of their community service programs.

The most common model of continuing professional development (CPD) that the government provides is "training of trainers." This process sees organizers inviting teachers to a training center for a certain period of time. On finishing the training, participants are expected to be trainers in their regions. The effectiveness of this model in practice is questionable for a number of reasons, including the variety of trainers, the different needs of the regions, and the availability of support services. Training for teachers is also on offer whenever the government implements a new policy or introduces new programs. For example, the government requires intensive training whenever it introduces a new curriculum. The government also provides scholarships for teachers wanting to study for higher formal qualifications.

Science Teacher Professional Associations

Indonesia hosts a number of teacher professional associations, but these are general teacher associations rather than science teacher associations. However, a science educator association called Perkumpulan Pendidik IPA Indonesia or PPII (Association of Indonesia Science Educators) came into existence at the time of writing (early in 2015). However, because it had just been founded, there was little to tell of its activity at this time. The association's aims are to improve the quality of its members and establish standards for science teachers. As mandated by law, certification programs must involve related professional associations so the roles of these associations have accordingly become crucial.

Teacher forums, popularly called MGPM (Musyawarah Guru Mata Pelajaran), also contribute to teachers' professional development. These forums usually include teachers who work in the same district or sub-district, and during these sessions teachers are expected to share their experience and to learn from one another. Teachers teaching in the same school subject are given one day off each week to attend a CDP session. If teacher forums for science teachers take place on Wednesday, then schools are not supposed to timetable science on that day so that the teachers can attend their local forum. In many cases, however, the number of MGMPs that actively conduct meetings is relatively small (Jalal et al. 2009). Many MGMPs do not schedule regular meetings, and when they do hold one, member participation is low. When a MGMP is first set up, the participation rate is usually relatively high but then it tapers off (Widodo and Riandi 2013).

Issues in Upper Secondary Science Teacher Quality

Teacher Supply

At one stage, the government brought in an edict to regulate teacher supply based on market demand. However, this move attracted objections from many universities, especially the private ones, because limiting the number of students meant less money from tuition fees. The lack of regulation on teacher supply has resulted in an oversupply of teachers, including science teachers. The new regulation that allows science graduates to become science teachers by taking professional teacher training makes the oversupply problem worse. As a result, numerous science teacher graduates have to find jobs other than their chosen career.

Most teachers who teach at government schools are civil servants, and teachers aspire to this status because it means security of employment. However, the number of teachers who achieve this status is small because it applies only to government schools. Many teachers who are not recruited by the government work either as part-time teachers or foundation teachers and are paid by the school itself or by the foundation. As independent bodies, private schools or foundations can recruit teachers according to their own criteria. As a consequence, the standards applied to recruitment tend to vary significantly.

Professional Education and Training Issues

Under the law, teachers are professionals who have undergone special education and training commensurate with this status. As noted earlier in this chapter, the government conducts a program that leads to professional certification for those teachers who successfully complete it. To enter the program, teachers must first pass an online competency test after which they attend a series of workshops, 90 hours in duration, run by appointed universities. The content of the workshops includes subject matter, pedagogy, classroom action research, and peer teaching. At the end of the workshops, teachers are evaluated to determine their eligibility for the certificate. Teachers awarded the professional certificate receive a special allowance from the government on top of their basic salary. The allowance effectively doubles their income.

The additional remuneration acts as a major incentive for teachers to obtain the certificate. However, because the law says that only teachers who hold a *sarjana* degree (four years of education at a university) can be considered for the certification program, many teachers have to continue their university studies first. Unfortunately, many of them do not take relevant education courses. Rather than choosing science education, for example, they might take subjects such as religious teaching or language. Therefore, while the teachers' level of educational attainment may rise, the skills and abilities needed in the classroom may not necessarily do so.

Issues at the Chalkface

In 2013, the Indonesian government launched a new curriculum (commonly called the 2013 curriculum). The 2013 curriculum introduced four core competencies—spiritual, social, content, and psychomotor. However, many teachers find these competencies difficult to understand. Spiritual competency and social competency are new additions to the national curriculum, and the intention behind their introduction is to increase people's belief in God while fostering their tolerance of religions other than their own, and to promote social cohesiveness and Indonesian unity. For most science teachers, these two competencies in particular are difficult to teach and assess.

Unlike the previous curriculum (commonly called KTSP) that gave regions and schools authority to develop their own programs, the new curriculum is a centralized one. Almost every aspect of the curriculum (school subject content, textbooks, lesson planning, and assessment) is centrally prescribed by the Ministry of Education. The rationale behind the adoption of a centralized curriculum is that of reducing education gaps between different parts of the country.

Implementation of the new curriculum has not been as smooth as anticipated, however. Lack of shared understanding of the nature and content of the curriculum amongst its developers as well as educational administrators and teachers has created confusion in the schools. For science, this difficulty is compounded by the limited availability of properly equipped science labs. Yet laboratory availability is a crucial success factor of the new curriculum because it prescribes a "scientific approach" wherein students learn the steps involved in scientific research through practical, hands-on science. Because of these issues, the government has decided to cancel implementation of the new curriculum for the time being and conduct further revisions.

One of the government's major efforts to improve science education quality was its program called RSBI ("pioneer international standard schools"). The program was based on the legal provision that requires central government and/or local government to operate at least one school that meets international standards at each educational level (Government of the Republic of Indonesia 2003). The term "international standard" was interpreted as invoking good levels of competency in mathematics and science as well as English. These three subjects were considered as the bases for developing competencies conducive to international market competitiveness. As part of its implementation of the program, the government ran bilingual classes at RSBI schools.

At first, schools and communities were very positive about the program. The communities saw RSBI as providing them with the means to deliver a better standard of education to their children. Parents applauded the label "international standard" and were willing to pay extra for their children to join international standard classes. But then the program began acquiring a negative image when some RSBI schools insisted that parents pay very high fees for their child's inclusion in the classes. RSBI schools began to be labeled as discriminatory because of the perception that only the children of the rich could attend them. Some RSBI schools also attracted criticism for failing to fulfill their promises of quality education, and indeed many schools that marketed themselves as RSBI did not show significant improvement in terms of their educational quality (Antara News 2010). Lessons continued to be conducted in the ways they had always been, and the academic performance of students who joined RSBI classes was usually no better than that of their peers in regular classes. In general, the price parents had to pay was not reflected by the high quality of education they had expected.

Based on a challenge submitted by a number of parents, on January 8, 2013, Indonesia's constitutional court published a judicial review that concluded that the establishment of RSBI had been illegal. As a result, the RSBI program ceased. From an academic perspective, one of the key factors that contributed to its failure was the quality of the teachers. Many mathematics and science teachers who taught at RSBI schools were incompetent to teach the subjects in English, but understandably so as they had not had the education they needed to teach the subjects in that language. The experience makes clear that educational initiatives must pay attention to teacher quality.

Continuing Professional Development Issues

Sustainability seems to be the most serious issue for CPD in Indonesia. In many cases, project-based CPD programs initiated by the government end when the pilot project that initiated them finishes. Even programs considered successful, such as PKG (Adey et al. 2004), have not been able to survive beyond the project phase. Another massive government-launched CPD program is "lesson study," considered to be one of Japan's key successes in improving Japanese education (Stigler and Hiebert 1999). Lesson study was introduced to Indonesia in 2004, and a number of reports suggest that it was having a positive impact on education quality (see, for example, Firman 2010; Hidayat and Jana 2011; Huzaeni 2010; Saito et al. 2006). Although lesson study is believed to be effective, Firman (2010) warns of the "wash-back" phenomenon that commonly arises out of initiatives of this kind. In particular, the popularity of lesson study seems to fade at the school level.

A survey carried out by Widodo and colleagues (2006) on the nature and impact of CPD programs in Indonesia identified a number of weaknesses in these programs. First, limited numbers of teachers actively participate in them. In many cases, the teachers taking part in the different programs on offer are actually the same teachers. Second, many of the programs do not sufficiently address practical issues. They are usually top–down in nature, with predetermined subjects, strategies, instructors, and time frames. As a result, they do not necessarily meet teachers' needs. Third, subject matter and methodology tend to be addressed as two separate domains: training in subject matter rarely involves the pedagogical aspects involved and vice versa. Teachers consequently have difficulty blending the two into a solid pedagogical content knowledge base. Fourth, because the CPD programs are conducted at training centers, teachers have to leave their school in order to participate. Having a teacher offsite is not a problem for a school if it can find another teacher to take over that teacher's class, but in those schools that cannot find a relieving teacher, students are typically left on their own.

Widodo and Riandi (2013) state that future CPD programs in Indonesia should adopt three strategies: providing participating teachers with incentives associated with clear career pathways; investing in the ICT and communication infrastructure that enables teachers to engage not only in face-to-face CPD but also in online CPD; and encouraging teacher education institutions to design inservice programs that prepare their graduates to be autonomous learners. Widodo and Riandi also claim that CPD models should be able to grow and change in keeping with the characteristics of the teaching cultures and educational communities in which they are implemented.

Although the scientific approach is a part of science itself, science teachers seem to find teaching science according to this approach difficult. Lack of understanding of scientific methods and lack of firsthand experience in conducting research are factors that hinder teachers in this respect. It is therefore reasonable to expect future CPD programs to provide teachers with firsthand experience of conducting research and to teach them how to convey the principles and practice of scientific research to their students.

Trends and Developments in Upper Secondary Science Teacher Quality

The issue of teacher quality and how to improve it has always drawn government attention. The government's teacher certification program described above was designed as a tool to improve teacher quality through three channels—the "attraction" channel, upgrading channel, and behavioral channel (Chang et al. 2014). The certification allowance has made teaching an attractive profession, such that teacher education is no longer a poor choice for high school graduates. The number of students enrolled in teacher education increased fivefold following the enactment of the pertinent law. However, it is this attraction that led to the present oversupply of teachers. Because teacher certification requires candidates to have a four-year university degree, underqualified teachers push themselves to upgrade. On the positive side, doubling teachers' income reduces teachers' financial concerns, thus helping them to focus on their jobs. Before the certification program, teachers often had one or two additional jobs.

A recent study on the impact of teacher certification suggests, however, that it does not lead to improved teaching performance or enhance teachers' professional behavior or participation in professional development (Chang et al. 2014). Another study showed that the pedagogical content knowledge of certified teachers is not necessarily better than the pedagogical content knowledge of student teachers in their final year of training (Hadiyanti et al. 2014).

Findings of studies on strategies to improve teacher quality indicate that a new alternative CPD paradigm is urgently needed. Most CPD models applied in Indonesia are taken from other countries that have different conditions (e.g., resources, geographical conditions, and culture). A model that works well in its country of origin may not work in other countries because of such differences. Limited teacher participation in online CPD sessions, for example, is not necessarily caused by a lack of the technology teachers need in order to study online but is more a product of them chatting on social media rather than using the technology for CPD purposes (Widodo and Riandi 2013).

Teacher certification programs aimed at improving teachers' competencies and teaching practice have largely turned out to be ineffective in Indonesia. Suggestions by Chang et al. (2014) deserve serious consideration. They consider the government should mandate certified teachers to spend a certain percentage of their professional allowance for continuing professional development activities and to seek recertification after a certain period of time. They also think that the government should establish and implement procedures for underperforming teachers. Implementing these suggestions may, however, be politically sensitive.

References

Adey, Philip Hewitt, with Gwen Hewitt, John Hewitt, and Nicolette Landau. 2004. *The Professional Development of Teachers: Practice and Theory*. Dordrecht: Kluwer Academic Publishers.

Antara News 2010. "RSBI Utuk Pres tasi Atau Gengsi," June 12, 2010, http://www.anta ranews.com/berita/1276342549/rsbi-untuk-prestasi-atau-gengsi (in Indonesian).

Central Bureau of Statistics. 2013. *Statistics Indonesia*. Jakarta: Author, http://www .bps.go.id.

Chang, Mae Chu, Sheldon Shaeffer, Samer Al-Samarrai, Andrew B. Ragatz, Joppe de Ree, and Ritchie Stevenson. 2014. *Teacher Reform in Indonesia: The Roles of Politics and Evidence in Policy Making*. Washington, DC: The World Bank.

Firman, H. 2010. "Dampak program kerjasama FPMIPA UPI dan JICA" [The impact of collaboration between FPMIPA UPI and JICA]. In *Teori, paradigma, prinsip dan pendekatan pembelajaran MIPA dalam konteks Indonesia* [Theory, paradigm, principle and approach of mathematics and science learning in Indonesian contexts], edited by T. Hidayat, I. Kaniawati, I. Suwarma, A. Setiabudi, and Suhendra, 41–54. Bandung: FPMIPA UPI.

Government of the Republic of Indonesia. 2003. *Undang-Undang Republik Indonesia No. 20 tahun 2003 tentang Sistem Pendidikan Nasional* [Law on the education system]. Jakarta: Author.

———. 2005. *Undang-Undang Republik Indonesia No. 14 tahun 2005 tentang Guru danDosen* [Law on teachers and lecturers]. Jakarta: Author.

Hadiyanti, Latifa Nur, Ari Widodo, Diana Rochintaniawati, and Riandi. 2014. "Pedagogical Content Knowledge of Experienced and Prospective Biology Teachers." Paper presented at the UPI-UPSI Conference, Bandung, Indonesia, June 25–26, 2014.

Hidayat, Arif S., and E. Jana. 2011. "Lesson Study and Improvement of Quality of Learning in Base Camp F. Sumedang District." In *Proceedings of the Fourth*

International Conference on Lesson Studies. Bandung: World Association of Lesson Studies.

Huzaeni, D. 2010. "Pelaksanaan lesson study berbasis sekolah (LSBS) di SMP Negeri Karawang Timur" [Implementation of school-based lesson study (LSBS) in junior high schools East Karawang]. In *Proceedings of the Third International Conference on Lesson Studies.* Bandung: World Association of Lesson Studies.

Indonesia University of Education. 2010. *Re-design pendidikan professional guru* [The re-design of professional teacher education]. Bandung: UPI Press.

Jalal, Fasli, Muchlas Samani, Mae Chu Chang, Ritchie Stevenson, Andrew B. Ragatz, and Siwage D. Negara, 2009. *Teacher Certification in Indonesia: A Strategy for Teacher Quality Improvement.* Jakarta: Ministry of National Education and World Bank, http://datatopics.worldbank.org/hnp/files/edstats/IDNprwp09c.pdf.

Ministry of Education and Culture. 2012. *Ikhtisar Data Pendidikan Tahun 2011/2012* [Summary of educational data 2011/2012]. Jakarta: Author.

National Education Standards Agency of Indonesia. 2009. *Standar Nasional Pendidikan* [National education standards]. Jakarta: Author, http://bsnp-indonesia.org/id/ (in Indonesian).

Saito, Eisuke, Imanssyah Harun, Isamu Kubokim, and Hiudeharu Tachibana. 2006. "Indonesian Lesson Study in Practice: Case Study of Indonesian Mathematics and Science Teacher Education Project." *Journal of In-Service Education* 32 (2): 171–84.

Stigler, James W., and James Hiebert. 1999. *The Teaching Gap: Best Ideas from the World's Teachers for Improving Education in the Classroom.* New York: The Free Press.

Suratno, Tatang, Ari Widodo, and Asep Kadarohman. 2013. "Interest and Recruitment in Science (IRIS) Study in Indonesia." In *Proceeding of the International Seminar on Mathematics, Science, and Computer Science Education 2013,* 1. Bandung: Faculty of Mathematics and University Education, Indonesia University of Education.

Widodo, Ari, and Riandi. 2013. "Dual-Mode Teacher Professional Development: Challenges and Revisioning Future TPD in Indonesia." *Teacher Development: An International Journal of Teachers' Professional Development* 17 (3): 380–92.

Widodo, Ari, Riandi, Amprasto, and A. R. Wulan. 2006. *Analyses of the Impact of Teachers' Professional Development Programs on the Improvement of Teachers' Teaching Practice: A Research Report on Educational Policy.* Bandung: Indonesia University of Education.

Further Reading

Chang, Mae Chu, Sheldon Shaeffer, Samer Al-Samarrai, Andrew B. Ragatz, Joppe de Ree, and Ritchie Stevenson. 2014. *Teacher Reform in Indonesia: The Roles of Politics and Evidence in Policy Making.* Washington, DC: The World Bank.

Hidayat, T., I. Kaniawati, I. R. Suwarma, Setiabudi, and Suhendra, eds. 2010. *Teori, paradigma, prinsip dan pendekatan pembelajaran MIPA dalam konteks Indonesia* [Theory, paradigm, principles and approaches to science teaching and learning in Indonesian contexts]. Bandung: FPMIPA UPI.

Jalal, Fasli, Muchlas Samani, Mae-Chu Chang, Ritchie Stevenson, Andrew B. Ragatz, and Siwage D. Negara 2009. *Teacher Certification in Indonesia: A Strategy for Teacher Quality Improvement.* Jakarta: Ministry of National Education.

Organisation for Economic Co-operation and Development/Asian Development Bank. 2015. *Education in Indonesia: Rising to the Challenge.* Paris: OECD Publishing, http://dx.doi.org/10.1787/9789264230750-en.

16

Sri Lanka

Marie Perera

Teaching Science in Sri Lanka

Backdrop

Known as the "Pearl of the Indian Ocean," the Democratic Socialist Republic of Sri Lanka (formerly Ceylon) is an island in the Indian Ocean, separated from southeast India by the Palk Strait. It has an area of 65,610 square kilometers. Sri Lanka's population of a little above 20 million is multi-ethnic and multi-religious. The 14 million Sinhalese comprise almost three-quarters of the country's population, while the next largest ethnic group, the Tamils, comprises 15 percent. The Moors and Malays make up 9 percent and just below 5 percent, respectively, of the population. Small minorities include Burgers or Eurasians, who are the people of mixed Asian and European ancestry. The main religious group is Buddhist at 70 percent of the population and is mainly Sinhalese. The Hindus are the next-largest religious group, comprising 13 percent of the population; they are mainly Tamils. Muslims and Christians are the other major religious groups at 13 percent and 10 percent, respectively. While most Muslims are Moors, the Christians are of mixed ethnicity.

The present education system in Sri Lanka is a blend of the legacy of its educational history and the impact of the global and local contexts. The earliest educational institutions in Sri Lanka were the Buddhist *pirivenas* (250 BCE to 1500 CE), which met both the religious and secular needs of the people. In addition to being able to attend the temple schools, some students went to the teacher's house (*gurugedera*) to learn. Unfortunately little information exists on the contemporaneous education of the Hindus and Muslims during this period. The guild system based on the teacher apprenticeship model offered vocational education. Transmission of skills from teacher to student contributed to the development of ancient Sri Lanka's amazing hydraulic technology and architectural marvels, some of which still exist in the ancient capitals.

The advent of the colonial powers from the sixteenth to the early twentieth centuries had an impact on the existing system of education. The Portuguese and the Dutch who controlled the coastal belt of the country established seminaries and parish schools, mainly for the purpose of propagating their religion. The Dutch also introduced the concept of compulsory education. The present education system was, however, mostly influenced by the British, who took control of the entire country in 1815. The main objective of their education system was to train personnel to assist in their administration. The main features of this education system were the dual (state and missionary) control of education and the consequent establishment of a dual system of schools: a minority of English-medium, fee-levying schools for the elite and the emerging middle class, and free "vernacular" schools in Sinhala and Tamil for the majority. The neglect of vocational education was also another notable feature of this system, as the need was to produce white-collar administrative assistance.

The introduction of universal franchise in 1931 resulted in redress for the disenfranchised masses. The new education policy as a result of the Kannangara reforms recognized the universal right to education by granting free education from Grades 1 to 13. Introduction of the mother tongue as the medium of instruction in the primary grades, with that medium to be gradually extended to the secondary and tertiary grades, was another aspect of these reforms. In addition, 54 "central schools" were established to extend modern secondary education, hitherto confined to urban centers, to the rural sector. These schools had facilities equivalent to the fee-levying English-medium schools. Every central school had a well-equipped laboratory, and students could learn science in both the junior secondary and upper secondary grades. Further, a Grade 5 scholarship scheme was introduced to enable talented students in village primary schools to access these schools. Localization of the curriculum and examinations and establishment of an autonomous university were also some of the policy changes that took place during the pre- independence era (1931–1948).

At present, the government plays a pivotal role in providing general education in Sri Lanka. The nation has approximately 10,400 schools, of which 91 percent are government schools. The balance consists of around 70 private schools, 700 *pirivena* schools (temple schools), and about 250 international schools. Total school enrolment is slightly in excess of 4 million students, of whom around 92 percent attend government schools. The government, private, and *pirivena* schools offer the national curriculum leading to the national public examination. International schools offer foreign curricula and prepare students for foreign examinations.

The government provides many welfare measures that have contributed to the significant gains in educational attainment in Sri Lanka. As well as providing education free of tuition costs from Grades 1 to 13 in public-sector schools, the government also provides a package of resources that includes free textbooks, free school uniforms, and subsidized transport. In addition, primary students in disadvantaged settings receive a free midday meal. The general education system is organized into three cycles: primary (Grades 1 to 5), junior secondary (Grades 6 to 9), and senior secondary (Grades 10 to 13). Children enter school at the age of 5 or 6 years. The government enforces legislation making it compulsory for all children of 6 to 14 years of age to complete 9 years of basic education. Most students continue up to Grade 11, but some discontinue their studies after the General Certificate of Education, Ordinary Level (GCE O/L) examination.

The school structure is divided into four types. Type 1AB schools are senior secondary schools that operate classes from Grades 1 to 13 or 6 to 13. These schools offer the General Certificate of Education, Advanced Level (GCE A/L) courses in three streams—science, commerce, and arts. These schools, of which there are 735, are considered to be the prestigious schools. Type 1C schools run classes from Grades 1 to 13 or 6 to 13. These schools, numbering approximately 2,040, offer GCE A/L courses in commerce and arts streams only. Type 2 schools have classes from Grades 1 to 11 or 6 to 11. There are 4,030 such schools in the system. Type 3 schools are primary schools that run from Grades 1 to 5 (or, occasionally, Grades 1 to 8). There are 3,125 such schools in Sri Lanka.

School Science

At present at the primary level, science is taught as an integrated subject under the Environment Related Activities (ERA) curriculum, to which six hours a week are allocated. The ERA syllabus is based on 16 themes relating to the social, biological, and physical aspects of the environment. Activity-based teaching in the primary grades is expected to achieve the identified competencies stated in the teacher guide (National Institute of Education 2009b). The medium of instruction in the primary grades for all subjects is the mother tongue. General science is introduced as a subject instead of ERA at the sixth grade. Education reforms in 1972 provided opportunities for all students from Grades 6 to 11 to learn science. At the secondary level, more emphasis is given to learning science through the objective of "science for all" so that students are able to apply theoretical knowledge to daily life. Physics, chemistry, and biology are taught as separate units in Grades 10 and 11.

The secondary curriculum is structured on five broad themes: observing the environment, organisms and life processes; matter, properties and interactions; Earth and space; energy, force, and work. Under these themes, the syllabus specifies eight competencies that are further divided into competency levels. The competencies define the expected learning while the competency levels relate to the teaching content. The syllabus has a spiral approach, which means the concepts are revisited at each grade with increasing depth. At the end of Grade 11, the first national examination, the GCE O/L, is held. However, in order to move to the next cycle, students do not have to gain a pass in the science subject unless they intend to enroll in the science stream.

The medium of instruction for teaching science is learners' mother tongue, except in bilingual classes. In 2003, the government introduced bilingual education to the lower secondary grades. This policy enabled schools with competent teachers to teach a few selected subjects in English, with the balance taught in the mother tongue. Science is one of the subjects taught in English. The purpose of this initiative was to provide students with the opportunity to acquire a level of English proficiency for higher education and employment purposes, while also improving subject knowledge. Although it was expected that bilingual education would gradually spread to all schools in the country, so far only 18 percent of schools have bilingual classes.

In the GCE A/L cycle, students undertake three subjects of study. They can select from physics, chemistry, biology, combined mathematics, higher mathematics, information technology, and agriculture. The medium of instruction is the mother tongue or English in bilingual classes. The GCE A/L examination, held at the end of Grade 13, has a dual purpose. It is not only a terminal achievement test with regard to secondary education but also the selection instrument for admission to the 14 national universities. Success on this examination is critical because of restricted entry to the national universities.

Although all students in the secondary grades have the opportunity to study science, such an opportunity is not available at the GCE A/L. The number of schools providing GCE A/L science subjects (type 1AB schools) is low and accounts for only 26 percent of the total number teaching GCE A/L subjects. Students, predominantly in rural areas, are deprived of the opportunity to study science at the higher level owing to non-availability of type 1AB schools in their locality.

Teacher Academic and Professional Education and Training

Sri Lanka has two entry points for people wanting to become science teachers: GCE A/L- qualified science graduates and BSc (Hons) or general degree holders. The first category includes those individuals who passed the GCE A/L examination but did not obtain the necessary cut-off mark to enter a university. They apply to a National College of Education (NCOE) that provides preservice training in science education. These trainees follow a three-year National Diploma in Teaching, at the end of which they obtain appointments as teachers of science. Six of the country's NCOEs are dedicated to providing preservice education in science teaching.

Degree graduates are given direct appointments as teachers. However, they must register in a Post-Graduate Diploma in Education (PGDE) program offered by a university or the National Institute of Education (NIE) within one year of securing a teaching appointment. The PGDE is a professional qualification that imparts pedagogical knowledge to all teachers irrespective of their discipline. The diploma program for science teachers has two components: methodology of teaching science and science education. During their teaching practicums, trainee teachers of science are expected to teach the subject of science in two secondary grades. The teaching practicum is considered the most important component of the PGDE courses.

In the full-time courses offered by the universities, teacher trainees are attached to schools in the vicinity of the universities for ten weeks. During this period, two university academics are assigned to each trainee as mentors. Six lessons (three each by the mentors) are evaluated, and trainees then receive feedback focused on improving their pedagogical skills. GCE A/L classes (Grades 12 and 13) in type 1AB schools and some 1C schools are set apart from Grades 6 to 11, and the graduate teachers who teach these classes teach these two grades only. Grades 10 and 11 classes are usually taught by A/L-qualified, professionally trained teachers from either a teacher training college or an NCOE.

Teacher training colleges provide training for non-graduates. The traditional three-year inservice teacher training program conducted by these colleges still prevails in two of Sri Lanka's teacher training colleges for teachers of science and mathematics. This program is for untrained teachers presently in the service as uncertificated teachers. However, the colleges are expected to follow a government-developed standardized core curriculum, the first two years of which are set apart for course work related to trainees' areas of specialization and general education courses. In their third year, trainees are posted to schools for the internship training. Once these colleges had cleared the backlog of untrained non-graduate teachers, the expectation was that they would then conduct inservice professional development for fully trained inservice teachers so that they could continuously review and update their knowledge, skills, and practices.

Until recently, the NIE conducted a distance-mode course leading to a BEd for untrained non-graduate teachers in the system. The course provided both content and pedagogical knowledge. However, this course has now been discontinued largely because of a lack of instructors, particularly outside the capital city Colombo. Today, the NIE offers only a general BEd, which it claims is suitable for both primary and secondary teachers.

Continuing Professional Development

Inservice teacher education programs in Sri Lanka can be broadly classified into two categories: long term and short term. Universities, the NIE, and teacher training colleges are the main providers of inservice teacher education in Sri Lanka. University programs focus mainly on teacher education for graduate teachers, while NIE, under the direction of the Ministry of Education, conducts teacher training programs for non-graduate teachers as well as graduate teachers. The inservice education provided by the universities generally serves to update science teachers' subject knowledge, and it takes the form of long-term courses leading to Master's degrees. The NIE also offers Master's degrees, but these lead to a Master's in education and are not subject specific. The NIE also provides opportunities for graduate teachers of science to upgrade their professional knowledge through short-term seminars and workshops. These short-term programs often relate to curriculum revisions.

Teacher centers attached to the NCOEs also provide continuing professional development for teachers. In the past, this development occurred through the "cascade model," whereby a few teachers from schools received training and were then expected to disseminate this knowledge at the school level. This model has been superseded.

Recently, school-based teacher development was introduced to the system as a pilot program under the Programme for School Improvement (Shrestha, cited in Policy and Planning Branch, Ministry of Education 2013, 30). The pilot involves 40 schools selected from 8 zones representing each province, and it relies heavily on the intervention of a group of specially trained officers who work at the zonal level with teachers according to the subjects they teach. Schools stipulate their own professional development priorities in line with school improvement planning.

Science Teacher Professional Associations

The two premier science professional associations in Sri Lanka are the Sri Lanka Association for the Advancement of Science (SLAAS) and the National Science Foundation (NSF). The SLAAS's science education committee is involved in school science projects, school science days, and Olympiads, all of which are intended to increase student interest in science. The association also conducts programs to build the capacity of both students and teachers. The Committee for Popularisation of Science of the SLAAS conduct a wide variety of programs annually in rural and urban schools in all districts of the country. These programs include seminars and lectures on scientific topics and quizzes conducted at the district level. Another popular activity among schoolchildren is the Nature Diaries program. Students engaged in this activity are encouraged to identify a sector, location, or activity in the natural environment, observe it over a period of several months, and maintain a diary on the changes they observe in the system. They then have to reason out the factors associated with such changes. Students who excel in these activities receive awards at the annual SLAAS sessions.

The NSF is a state-funded institution under the Ministry of Technology and Research. The foundation supports research in all fields of fundamental and applied sciences as well as in the social sciences. The foundation's School Science Societies program aims to foster students' acquisition of knowledge on the latest developments in various fields of science and technology and to increase students' awareness of the application of scientific knowledge in day-to-day activities as well as in improved living standards.

The NSF also funds resource persons for scientific programs organized by the school science societies and school science projects based on proposals submitted. It furthermore runs a scheme for publishing science books and booklets, with the aims of increasing the availability of printed material in science and technology relevant to local needs and encouraging potential writers in science and technology. Another aim of the scheme is to improve the quality of publications on science and technology and to create an avenue for the public to obtain authentic reading material in these two areas.

Another NSF school-related initiative is the Vidunetha program, which the foundation launched in order to guide schoolchildren on how to conduct a research project using scientific methodology and to motivate them to conduct such projects. More specifically, the program aims to teach scientific concepts and inculcate scientific thinking and creativity among schoolchildren. The target group for this program is students at the junior secondary level (Grades 6 to 9) and their teachers. As part of the program, the foundation conducts regional workshops in collaboration with the provincial coordinators of the Ministry of Education. The foundation is also committed to facilitating awareness of the value of science among various sectors of the community, such as industrialists, teachers, scientists/researchers, media personnel, and the general public.

Student involvement in these various NSF programs depends on their teachers' interest in them. The programs have yet to reach all schools, but the number of schools participating in them has increased over the years. This increase is a positive development, as it results in increased interest in the practical aspects of science.

Issues in Upper Secondary Science Teacher Quality

Teacher Supply

In Sri Lanka, responsibility for ensuring the education system has all the teachers it needs lies with the central government. Each year, the government recruits the number of teachers needed to accommodate increases in student enrollment and to compensate for teacher attrition. The number to be recruited each year is the difference between teacher requirements and the total number of teachers available.

The practice in Sri Lanka of classifying schools according to the facilities available and their location also determines teacher recruitment and deployment. Classification is based on ten criteria, such as accessibility of transport, basic facilities such as availability of electricity and water supplies, and school environment ("very congenial," "congenial," "difficult," and "very difficult"). All teachers in Sri Lanka are mandated to serve in difficult or very difficult schools for a minimum period of three years. However, teachers frequently use political or other forms of influence to avoid serving in such schools. Even if teachers accept appointments to these disadvantaged locations, they often transfer to a more congenial locality before their mandatory service period is over. Low teacher salaries and a lack of incentives hinder deployment to difficult areas, and teachers prefer to live and work in cities, towns, and prosperous urban areas. These preferences lead to overstaffing at urban schools and understaffing in rural areas, with adverse effects on learning outcomes in disadvantaged rural communities.

Successive governments have used public-sector recruitment to create employment opportunities for educated youth, including university graduates aspiring to take up government jobs. Many of these educated youth, particularly university graduates, are absorbed into the teaching service, as the network of 10,000 schools provides the government with plenty of destinations to which employees can be assigned. But there is a chronic mismatch between teacher demand and supply. Although the NCOEs and universities provide most of the teachers to the system, they are not adequate to meet actual teacher requirements. Therefore, the last few decades have seen subject-specific teacher shortages, and science is one such subject. Not surprisingly, such shortages are more acute in remote schools. As one of its measures to redress this issue, the government periodically appoints unemployed graduates or A/L-qualified young people as science teachers. However, this approach results in a number of untrained science teachers always evident in the system.

According to Ministry of Education statistics, four provinces have 100 percent or more available science/mathematics teachers for the secondary grades. However, this does not mean the required number of teachers is available in all schools in these provinces because often the disadvantaged schools, even in these provinces, have a dearth of science/mathematics teachers. In contrast, in the other five provinces, the required number of teachers overall is less than the available number of teachers. Hence, the student–teacher ratio also varies from province to province and between schools as well.

The supply of teachers also varies according to the medium of instruction. While overall there is an excess of Sinhala-medium teachers, there is a deficit in Tamil-medium and bilingual schools. One of the reasons for the slow spread of bilingual education in Sri Lanka is the lack of teachers to teach science/mathematics in the English medium. Although the medium of instruction in the faculties of science is English, most undergraduates have studied in the medium of their first language at upper secondary level. Furthermore, due to the lack of teachers in certain schools, some have not even studied English as a subject. Hence, even though at graduation these teachers have subject competency, their language competency is low. Those who are competent in English can moreover obtain paid better employment in the private sector.

Professional Education and Training Issues

Sri Lanka does not have an organization for regulating the standardization and quality assurance of the teacher education programs offered by different institutions, nor is there coordination between the institutions that provide teacher education. This situation has resulted in several shortcomings associated with the system of teacher education. Because standards are not stipulated, program quality varies across the various institutions even though they offer the same qualifications or level of certification.

As noted earlier, six NCOEs provide preservice education to prospective science/mathematics non-graduate teachers destined for Grades 6 to 11 teaching. Teacher educators at the NCOEs are expected to possess high academic and professional qualifications. However, adequate numbers of such personnel are generally not available, so temporary staff with lesser qualifications are brought in to fill the gap. However, ongoing lack of fully qualified staff means that in most of the NCOEs, these temporary lecturers continue working well beyond the period intended, thus compromising the overall quality of teaching in these colleges.

For graduates who enter the teaching profession, preservice training is provided by universities and NIE. The conventional universities offer the PGDE on campus. The Open University of Sri Lanka trains teachers through distance learning. Both the Open University and the NIE offer face-to-face components of their courses in the regional centers. The majority of the teacher educators who teach in the regional centers are not university academics, and there are allegations that the quality of teaching in these places is low. Generally, there is more demand for courses in the universities.

Because acquisition of the PGDE is mandatory for entry to a teaching post, the demand for places in the PGDE courses outstrips what the institutions can accommodate. In an effort to cater for this demand, the conventional universities also offer part-time courses during the weekends in addition to their full-time regular courses. During the part-time courses, trainees engage in practicums for ten weeks in their own school. Due to the large numbers involved, "master teachers"—teachers who are meant to be highly experienced and qualified "experts"—serve as mentors for these trainees. Since there are no benchmarks or standards by which to evaluate the competencies of these master teachers, the quality of their mentoring role is questionable.

Issues at the Chalkface

Practical work, which should be an important component of any science curriculum, receives very little attention in Sri Lanka's secondary curriculum (World Bank 2011b). According to the teacher instructional manual for Grade 6, which gives an introduction to the Grades 6 to 11 science curriculum (see, for example, National Institute of Education 2007), the objectives of the science program are to help students develop and apply scientific knowledge and concepts to everyday living and to the nation's well-being through the use of inquiry skills, problem solving, and scientific reasoning. However, commentators argue that the activities relating to the competency levels identified are directed more toward learning content and less toward promoting student-directed inquiry (see, for example, McCaul 2007). Research findings published by the National Education Research and Evaluation Centre (2005, 2008, 2013) reinforce these claims.

A curriculum revision for the secondary grades took place in 2007, after which a 5E learning cycle of engagement, exploration, explanation, elaboration, and evaluation was introduced with the intention of making learning more effective (Ginige 2008). According to several studies, implementation of this method has been problematic. Some of the reasons cited are time constraints associated with a curriculum that is too content heavy, teachers' lack of awareness of the new methodology resulting in them reverting to the lecture method, and inadequate training opportunities for teachers (McCaul 2007; National Institute of Education 2008, 2009a).

The syllabuses for the GCE A/L science subjects were also put onto a competency-based footing from 2012 onward, with the concurrent expectation that science learning would be student centered and activity oriented. GCE A/L teachers now have greater freedom to follow the teaching/learning method of their choice in order to achieve the relevant learning outcomes (National Institute of Education 2010). However, many of the issues associated with the junior secondary and GCE O/L science curriculum are also common to the GCE A/L. As documented (World Bank 2011b), teachers and provincial education authorities claim that content overload, lack of emphasis on learner-centered teaching, teacher shortage and teacher quality, neglect of practical work, and insufficient laboratory facilities are issues equally applicable to both junior secondary and upper secondary science teaching.

Several reasons for the neglect of practical skills can be identified. One is that these skills are not assessed in the public examinations. In a highly competitive education system, what is tested in the examinations has a washback effect on what is taught in the classroom. Another reason is the lack of laboratory resources in the schools. Introduction of bilingual education has worsened this situation because preference with regard to laboratory access is given to monolingual classes, and these classes contain the higher numbers of students. Another reason is teachers' lack of confidence or skill in handling practical work, which is understandable, as this is an area that is neglected in continuous professional development.

The public examinations, moreover, do not test the competencies and learning outcomes delineated in the new syllabuses. Research findings reveal that neither the GCE O/L examination nor the GCE A/L adjusted the evaluation format to test all the competencies identified and so still placed more emphasis on cognitive abilities, as in the past. Consequently, teachers too have not changed their assessment practices at the school level, and there are also problems with papers prepared by the zonal education departments (McCaul 2007; National Institute of Education 2008, 2009a)

Problems in the curriculum affect the quality of teaching. Teachers complain of the over- burdened curriculum for the secondary grades and say they find it difficult to cover the syllabus within the stipulated time. Students, according to teachers, are also unable to cope with the large volume of scientific knowledge they have to learn. Teacher instructional manuals sometimes restrict teachers' use of innovative strategies in teaching. Hence, they claim that neither teaching nor learning science is an enjoyable activity (World Bank 2011b). There are also instances where learning materials are not ordered in the proper sequence, such as an application of a principle preceding the learning of the concept!

Continuing Professional Development Issues

As noted earlier, a number of professionally unqualified teachers can still be found in Sri Lanka's education system. Although facilities providing professional training are available, they are not fully utilized. There is no mechanism to monitor the professional competency of teachers and to ensure that all teachers receive training. Some untrained teachers are reluctant to enter a teachers college to undergo institutional training because they are usually deployed to "difficult" area schools after the training.

As already mentioned, the teachers' colleges provide non-graduates with long-term inservice training. The backlog of untrained non-graduate teachers was expected to clear shortly after the NCOEs began providing preservice training to teacher recruits in 1985. Teachers colleges were then required to conduct continuous professional development (CPD) courses for trained teachers. However, because untrained non-graduates have continued to be part of the teaching force, two teachers colleges are still, with respect to science teaching, providing a three-year inservice training program leading to a Certificate in Teaching Science. These teachers are expected to teach science/mathematics in Grades 6 to 11 while science graduates are expected to teach Grades 12 to 13. However, in schools where there are teacher shortages, principals sometimes ask NCOE diploma holders to teach students in Grades 12 to 13 as well.

At the provincial level, CPD is meant to be further disseminated through a network of inservice advisors (ISAs), but according to a study conducted by Gunawardhana (2011) on short-term CPD, stakeholders are not very happy with the courses offered by both the NIE and at provincial level. According to this study, specifically stated negative aspects of the training in science and mathematics include shortcomings in the use of new technology in inservice sessions and some ISAs not conducting demonstration lessons. The situation is worse in the sessions for bilingual teachers, according to the teachers, who claim that the quality of the English-medium science sessions is not up to standard. Practical sessions are seen as not being methodical. ISAs visit the schools only on the day of the zonal supervision. Hence, the ISA's visit becomes more of an inspection than a mentoring occasion. Another problem noted by teachers is a lack of ISAs for science in some zones. According to some teachers, there are qualified teachers who have passed the competitive selection examination and so could become ISAs, but still the vacancies exist. Using information and communication technologies and conducting practical sessions and activity-based sessions are among teachers' suggestions for improving inservice training relating to science subjects.

NIE does provide short-term training, especially when curriculum revisions take place and teachers need to be made aware of the changes. This training uses the cascade model wherein ISAs or one teacher from each school is trained first in the expectation that they will then train their colleagues or the teachers within their zones. However, it has been claimed that this approach has severe shortcomings because the trained are often unable to transmit the knowledge effectively to the next level of trainees (World Bank 2011a). In 2010 an attempt was made to conduct "direct training" involving NIE staff and teacher trainers in the provinces. The hope was that this method would see all teachers in the provinces receiving their training directly from NIE personnel or an ISA. However, it seems that the provinces experience considerable difficulty in conducting such training due to lack of funds and deficient training facilities in these areas (World Bank 2011a).

Trends and Developments in Upper Secondary Science Teacher Quality

An initiative currently underway is development of a network of 1,000 good-quality secondary schools. Networks of schools made up firstly of secondary schools and feeder primary schools situated close to those secondary schools have been selected through a mapping exercise carried out via a determinant-based methodology. Six hundred and thirty-two secondary schools and 2,000 primary feeder schools connected to those schools have been selected for development, and 1,000 science and mathematics teachers are to be appointed to the schools under a program called the One Thousand Schools Programme.

As discussed earlier, one of the problems in Sri Lanka with respect to science teacher quality is the recruitment and deployment of teachers. One of the suggested policy options for addressing this issue is that of moving away from the centralized teacher employment system and establishing a decentralized one in which schools recruit teachers directly. Of the 1,000 science and mathematics teachers to be recruited under the One Thousand Schools Programme, 485 science graduates had been recruited as of 2014 (Ministry of Education 2014). It is hoped that this policy will lead to a paradigm shift in teacher recruitment and deployment. Another issue with regard to teacher quality is retaining teachers in the service, especially in rural and disadvantaged areas. Policies currently under consideration with respect to this issue involve the provision of fiscal and other incentives, such as faster promotion, for teachers serving in schools located in the disadvantaged areas.

CPD is another issue that needs urgent attention. The Ministry of Education has accordingly begun training 240 officers in charge of science in collaboration with the SLAASs. In addition, provincial ministries of education have started establishing teams of provincial resource persons and training 350 officers in 7 provinces (Ministry of Education 2014). However, the Ministry also needs to formulate an overall policy framework that sets goals for science education throughout Sri Lanka and delineates pathways enabling teachers to teach in ways that will achieve those goals. Such a framework would be beneficial to curriculum developers, planners, and implementers. There is also the need for a regulating body for standardization and quality assurance of the teacher education programs offered by different institutions. This development is a must if Sri Lanka is to maintain teacher quality.

References

Ginige, I. L. 2008. *The Vision of the New Curriculum and the Paradigm Shift in School Education.* Kurunegala: Vocational Skills Development Centre.

Gunawardhane, Raja. 2011. "A Study on the Effectiveness of Short-Term In-service Teacher Training in the Teaching–Learning Process." Nugegoda: National Education Commission, http://nec.gov.lk/wp-content/uploads/2014/04/inservicett.pdf.

McCaul, T. 2007. *Study of the Implementation of Mathematics and Science Curriculum in Grades 6 and 10.* Maharagama: National Institute of Education.

Ministry of Education. 2014. "Sri Lanka Education Information 2014," http://www.moe.lk/.

National Education Research and Evaluation Centre. 2005. *National Assessment of Achievement of Grade 8 & 10 Students in Sri Lanka.* Colombo: NEREC, University of Colombo.

———. 2008. *National Assessment of Achievement of Grade 8 Students in Sri Lanka.* Colombo: NEREC, University of Colombo.

———. 2013. *National Assessment of Achievement of Grade 8 Students in Sri Lanka.* Colombo: NEREC, University of Colombo.

National Institute of Education. 2007. *Science Grade 6: Teacher Instructional Manual.* Maharagama: Author.

———. 2008. *An Evaluation of the Process of Development and Implementation of the New Curriculum in Grades 6 and 10.* Maharagama: Author.

———. 2009a. *An Evaluation of the Process of Development and Implementation of the New Curriculum in Grades 7 and 11.* Maharagama: Author.

———. 2009b. *Primary Curriculum.* Maharagama: Author, http://www.nie.sch.lk/ebook/s5tim35.pdf (in Sinhala).

———. 2010. *GCE Advanced Level, Biology Grade 13 Teacher's Instructional Manual.* Maharagama: Author.

Policy and Planning Branch, Ministry of Education. 2013. *Education Sector Development Framework and Programme (ESDFP) (2012–2016).* Battaramulla: Author, http://www.moe.gov.lk/english/images/publications/ESDFP2012-2016/English ESDFP.pdf.

World Bank. 2011a. *Transforming School Education in Sri Lanka: From Cut Stones to Polished Jewels.* Colombo: Author, www-wds.worldbank.org/external/default /WDSContentServer/WDSP/IB/2011/12/16/000386194_20111216003147/Ren dered/PDF/660360PUB00PUB0Report0final0version.pdf.

World Bank. 2011b. *Strengthening Science Education in Sri Lanka.* Colombo: Author, http://www-wds.worldbank.org/external/default/WDSContentServer/WDSP/IB /2011/12/19/000333038_2011121922.

Further Reading

Aturupane, Harsha. 2009. *The Pearl of Great Price: Achieving Equitable Access to Primary and Secondary Education and Enhancing Learning in Sri Lanka.* Brighton: University of Sussex.

Azeez, A. M. A. 1969. "The Muslim Tradition." In *Education in Ceylon (from the Sixth Century B.C. to the Present Day): A Centenary Volume*, vol. 3, edited by U. D. I. Sirisena, 1145–57. Colombo: Ministry of Education and Cultural Affairs.

Department of Examinations. 2012. *National Symposium on Reviewing the Performance of School Candidates, GCE(O.L) Examination–2011*. Battaramulla: Department of Examinations.

Guruge, A. 1969. "Science and Technology." In *Education in Ceylon (from the Sixth Century B.C. to the Present Day): A Centenary Volume*, vol. 3, edited by U. D. I. Sirisena, 91–100. Colombo: Ministry of Education and Cultural Affairs.

Hevawasam, P. 1969. "The Buddhist Tradition." In *Education in Ceylon (from the Sixth Century B.C. to the Present Day): A Centenary Volume*, vol. 3, edited by U. D. I. Sirisena, 1107–30. Colombo: Ministry of Education and Cultural Affairs.

Ministry of Education. 2010. *School Census: 2010*. Colombo: Author.

Ministry of Higher Education. 2012. *Sri Lanka Qualifications Framework*. Colombo: Author, http://www.ugc.ac.lk/attachments/1156_Sri_Lanka_Qualifications _Framework.pdf.

National Education Commission. 1997. *Reforms in General Education*. Nugegoda: Author, http://nec.gov.lk/post4-2/.

———. 2003. *Envisioning Education for Human Development Proposals for a National Framework on General Education in Sri Lanka*. Colombo: Author, http://nec.gov .lk/post4-2/.

National Science and Technology Commission. 2008. *National Science and Technology Policy*. Colombo: Author.

Perera, G. M. T. N. 2008. *An Evaluation of the Process of Development and Implementation of the Curriculum in Grades 6 and 10*. Maharagama: National Institute of Education, http://www.nie.sch.lk/resource/eva610en.pdf.

Policy and Planning Branch, Ministry of Education. 2013. *"School First": Education Sector Development Framework and Programme (ESDFP) (2013–2017)*. Battaramulla: Author, http://planipolis.iiep.unesco.org/upload/Sri%20Lanka/Sri_Lanka _ESDFP_2013-2017.pdf.

Ruberu, Tantrige Ranjith. 1969. *Education in Colonial Ceylon: Being a Research Study on the History of Education in Ceylon for the Period of 1796 to 1834*. Kandy: Kandy Printers.

Russell, R. Ross, and Andrea Matles Savada, eds. 1988. *Sri Lanka: A Country Study*. Washington, DC: GPO for the Library of Congress.

Sethunga, Prasad, R. D. Sugathapala, and J. Jayasuriya. 2007. *The Study on School Based Teacher Development Programs*. Peradeniya: Department of Education, University of Peradeniya, http://www.pdn.ac.lk/arts/education/images/SBTD%20FINAL%20 REPORT.pdf.

Somasegaram, S. U. 1969. "The Hindu Tradition." In *Education in Ceylon (from the Sixth Century B.C. to the Present Day): A Centenary Volume*, vol. 3, edited by U. D. I. Sirisena, 1131–43. Colombo: Ministry of Education and Cultural Affairs.

"Sri Lanka Association for the Advancement of Science (SLAAS)," http://www.slaas .lk/.

University Grants Commission Sri Lanka. 2014. *Admission to Undergraduate Courses of the Universities in Sri Lanka, Academic Year 2013/2014*. Colombo: Author, http://www.ugc.ac.lk/downloads/admissions/local_students/Admission%20 to%20Undergraduate%20Courses%20of%20the%20Universities%20in%20Sri%20 Lanka%202013_2014.pdf.

17

Cluster Summary Two

The countries making up this cluster span a very large national income range, with an order of magnitude separating the highest from the lowest. Oman is on a par with South Korea in this regard but presents us with a rapidly evolving system that has more in common with many poorer countries than with the comparatively settled education systems characterizing the members of the first cluster. Conversely, Turkey with its long-established, prestigious education system arguably has more in common with countries in the first cluster, other than its categorization as middle income.

Teacher Supply

Reliance on expatriate secondary school teaching personnel remains high in Nauru and Oman and was, until recently, a historically traditional feature of the Fijian and Malaysian education systems. Teaching as a career does not generally appear to have the same allure in this second cluster of countries, where it tends to be relatively poorly remunerated, although it does offer employment stability. Most of the countries under consideration have little problem meeting teacher quotas in general. However, shortages in some subject fields, including upper secondary science, may occur, as they do in the more affluent cluster.

Teacher Academic and Professional Education and Training

In general, upper secondary science teachers in this cluster are in possession of tertiary qualifications in science and have undergone pedagogical training, although some poorly qualified and/or untrained teachers may remain in the system. These people are, however, given the opportunity to upgrade their qualifications through work/study arrangements. There is a trend toward a greater degree of uniformity across the multiplicity of teacher training providers, with that trend prompted by government authorities setting course content and delivery standards.

The provision of continuing professional development (CPD) appears to be clearly associated with national income, with the pattern being one of more CPD in terms of both quantity and quality where per capita GDP is higher. As we move down the national income ladder, CPD appears to become more dispersed and lacking in focus, as well as relying more on models that involve training a handful of teachers at the local level who are then supposed to act as further disseminators of that training.

With the growing realization that professional education is a career-long process, some systems are making genuine efforts to raise the profile and effectiveness of their CPD provision, but the success of these efforts requires a fiscal base that the less well-off countries have difficulty providing.

Issues at the Chalkface

Success in high-stakes terminating examinations is the principal objective of most upper secondary students in these systems, given that competition for places in sought-after university programs is often fierce. It is probably fair to say that a "good teacher" at the upper secondary level in these systems is one whose students score highly on the "big exam." Thus, it is little wonder that teaching tends to be conventional and content focused. An opposing pressure comes from revised curricula that incorporate ideological imports pertaining to constructivist teaching and learning methods. Teachers are then faced with the quandary of having to serve two masters. In practice, they serve one (student and parental expectations relating to examination preparation) and pay lip service to the other.

In general, upper secondary teachers in this cluster have available to them the physical resource base that they need to do their jobs, although this may be stretched in the case of the least affluent countries. Ironically, the emphasis on examination preparation often reduces the profile of practical work to the extent that few physical resources are required. The broadening of terminating assessment to incorporate school-based practical work and the growing pressure on teachers to use inquiry-based approaches place additional demands on the resource base.

Science Teacher Professional Associations

With the exception of Turkey and, as of late, Indonesia, upper secondary science teachers in this cluster do not have a national professional science teacher association that they can join. However, various other associations, both education based and science based, are available to teachers in some countries.

18

Yemen

Saouma BouJaoude and Khalid Khan

Teaching Science in Yemen

Backdrop

Yemen is located in southwest Asia at the southern border of Saudi Arabia. The Republic of Yemen is one of the poorest Arab countries and faces a wide range of developmental challenges. The country has one of the highest population growth rates globally, and the population is expected to double in the next 23 years to around 40 million. As a result, the demand for educational and health services, drinking water, and employment opportunities will increase. Social development indicators such as child malnutrition, maternal mortality, and educational attainment remain discouraging. There are also large gender discrepancies: women have poor access to economic, social, and political opportunities.

Before 1990, Yemen was two independent countries—North Yemen (Yemen Arab Republic) and South Yemen (People's Democratic Republic of Yemen), each with its own education system. These two countries merged in 1990 to become the Republic of Yemen, at which time all government ministries, including education, also merged.

Before 1962, the year the Yemen Arab Republic was established, North Yemen was a largely closed society, and education was limited to religious schools where children memorized the Koran. Schools were set up under local initiatives. However, not all children had access to these schools. The majority of students were boys; very few girls attended school. The North Yemen government's efforts to reform education started in 1962. The first Ministry of Education was established in 1963 to monitor the public school system. The foundations of the school and post-school education system were established between 1963 and 1970. This was followed by the first five-year plan (1970–1975), during which primary education became universal, Sana'a

University was founded, and the technical and vocational education system was established.

During the1970s and the 1980s, Yemen saw an expansion of primary education for males and females, attempts to improve the quality of education at all levels, and establishment of informal education programs, especially those related to adult literacy. Additionally, attention was paid to linking education to the needs of society and the labor market, attracting qualified individuals to the teaching profession, and improving the quality of teachers at all levels. Throughout this period, religious schools continued to operate, and a few private schools opened. The school system included six years of primary, three years of lower secondary/middle, and three years of upper secondary schooling. Students at the upper secondary level were tracked into academic (scientific and literary), vocational, commercial, agricultural, or teacher training streams.

The expansion of education in the People's Democratic Republic of Yemen (South Yemen) began in1967 after the British withdrawal. During the time of British rule, education was available only in Aden, the capital. Primary and preparatory schools existed in a few neighborhoods in Aden, while only one girls' secondary school and two private schools existed in other neighborhoods of the city.

During the 1970s, several education plans were developed for the new republic, and the educational landscape of South Yemen had many common features with North Yemen. South Yemen also adopted the 6-3-3 education system. However, in 1975 it changed to an 8-4 structure (eight years basic schooling and four years of secondary school). Secondary education students were tracked into academic, vocational, and technical or teacher training education. The education system in South Yemen was tuition-free for all students. In addition, all students received free textbooks. Students in those rural areas with no schools received free board to attend school in areas where there were schools.

After the unification of Yemen, the two education systems merged into a single system of nine years of basic education and three years of upper secondary education. A Ministry of Education and Teaching (MOE) was established in 1992 to monitor the system (MOE 1992).

Today, students who are tracked into the academic stream after Grade 9 follow a common curriculum in Grade 10, after which they are tracked into scientific and literary streams. Entry to the upper secondary level is determined by success on a national examination at the Grade 9 level. Students who pass the examination receive a Basic Education Certificate. After Grade 9, students can move to a technical secondary school, a vocational training center, a health manpower training institute, or an agricultural secondary school according to their needs. The school system thus caters for a variety of different skills at different levels in accordance with the national quest to build up a pool of skilled labor. At the end of Grade 12, all secondary students take a national examination. If they pass it, they are awarded the Secondary Education Certificate.

According to Yemen's Education Policy and Data Center (2014), the primary education gross enrollment rate in Yemen is 97 percent, which decreases to 57 percent at lower secondary level, with a student transition to secondary school of 83 percent. Transition to university is controlled by the awarding of university admission status based on required levels of achievement in the Secondary Education Certificate.

Enrollment at the secondary level in Yemen has increased significantly over the past three decades. According to a World Bank report published in 2010, enrolment in secondary education is better than in other low-income countries, even though it is significantly lower than elsewhere in the MENA region (Middle East and North Africa) and many other countries throughout the world. In the past decade, female enrollment has increased while that of males has decreased. However, the enrollment rate for females and in rural areas is still lower than that of males and in urban areas. According to UNICEF data, the secondary school participation rate for males in 2012 was close to 50 percent while that for females was about 30 percent.

Even though most Yemeni students attend public-sector schools, the private sector is playing a growing role in education. According to the Yemen Education Sector Plan for the period 2013 to 2015 (MOE 2013), the private sector covers 2 percent of basic and secondary enrollments. However, Yemen seems to be moving toward more private-sector involvement in educational provision as evidenced by the fact that one of the MOE's priorities is to encourage the private sector to invest in education at all levels (MOE 2013).

The directorate of Accreditation and Quality Assurance of schools was established at the beginning of 2012 with the purpose of improving educational quality at the pre-tertiary level. Since then, the MOE has established quality assurance offices in all governorates, signed an agreement with UNICEF to develop quality assurance guidelines and guidebooks, published a comprehensive framework for collecting evidence for quality assurance (including the functions and responsibilities of MOE staff), and published an accreditation standards handbook for experimental modernized schools and regular public schools. The MOE has furthermore facilitated a culture of quality by organizing training sessions for staff of the governorate quality-assurance offices and also for governorate education directors, implementing educational awareness lectures for community organizations, and establishing partnerships with Yemeni universities toward this end.

School Science

There is a dearth of research information on science education in Yemen in either English or Arabic. The principal sources of information are reports prepared by international organizations such as UNESCO and the World Bank in cooperation with Yemen's Ministry of Education and Teaching.

At present, Yemen has a compulsory school curriculum from Grades 1 to 9 that includes six essential learning areas, including science. In these grades, science is taught in an integrated manner, even though the topics are divided into biology, chemistry, physics, and Earth and space science. At the secondary level, "biology and environment," chemistry, physics, and Earth and space science are separate subjects. Biology, environmental science, and geology are presented in the same book and taught by the same teacher. A new Yemeni science curriculum, among other subjects, was implemented in Grades 1 to 6 in 2000, followed by a new curriculum for Grades 7 to 9 in 2001 and for Grades 10 to 12 in 2003.

Science in Grades 1 to 6 is divided into seven major themes: biodiversity, human body and health, the Earth, the atmosphere, galaxies and the solar system, matter, and energy, force, and motion. At the preparatory level (Grades 7 to 9), there are nine themes: matter, energy, force and motion, biodiversity, human body and health, life processes, Earth, weather and the atmosphere, and galaxies and the solar system. In Grade 10, which is common to all students, ten major themes revolve around biodiversity, living processes, the environment, Earth, force and motion, matter, energy, electronics, astronomy, and general, inorganic, and organic chemistry. The distribution of the themes in Grades 11 and 12 is the same as that of Grade 10, except that Earth and space science are somewhat downplayed in Grade 11.

The objectives of science at the upper secondary level according to the MOE include deepening learners' Islamic faith, reinforcing their understanding of science content taken during the preparatory stage, equipping them with the knowledge and skills necessary to pursue higher education or join the workforce, and helping them to acquire scientific facts, understand current developments in knowledge, and solve academic and everyday problems. Moreover, education should increase learners' awareness of the appropriate use of technological and scientific innovations, help them to develop healthy habits, protect the environment, conserve natural resources, and enhance their appreciation of the pioneering work of Arab and Muslim scientists in all realms of science.

The number of periods allocated to science makes up a relatively low percentage of the total number of periods taken by students at the basic education level. This number increases with grade level to reach around one-quarter of the number of periods in Grades 11 and 12 of the scientific stream. Students in the literary stream do not take any science subjects.

The MOE produces good-quality textbooks and teachers' guides (World Bank 2010). These are available on the ministry's website. The ministry has also produced a laboratory manual. It, too, is available on the ministry's website.

Data published by MOE in 2008 showed that most students had a preference for the science stream: in 2007/2008 approximately 80 percent of the students ended up in it. However, only a minority eventually pursue science-related majors at the university level (MOE 2008). According to UNESCO (2011), the major reason why students prefer the science stream is that it allows them to apply for more majors at the university level if they successfully complete this stream.

Teacher Academic and Professional Education and Training

Teachers in Yemen are predominantly male and generally unqualified. MOE officials and international organizations claim that almost half of Yemen's teachers lack the qualifications they need to be effective. Approximately 65 percent of these teachers do not have Bachelor's degrees (Al-Sakkaf 2014; World Bank 2010). The minimum educational qualification for teachers is a two-year post-secondary diploma as stipulated by the MOE in 1998. This diploma was offered in two-year teacher training institutes whose role changed as of 2000 when they became centers of professional development because of the increase in the number of universities offering teacher training programs. In practice, teachers who receive this diploma teach Grades 1 to 6 whereas university graduates are the preferred group for teaching at the preparatory (Grades 7 to 9) and secondary (Grades 10 to 12) levels because they are viewed as subject-matter specialists (UNESCO 2011).

As was the case with the education system in general, two systems of teacher education existed before unification in 1990. However, preservice and inservice teacher education became one after unification. According to Al-Hage (2006), many preservice and inservice programs are presently offered by the 10 public colleges of education and the 31 public teacher training institutes. The objectives of these facilities include ensuring professional teachers have strong content and pedagogical knowledge and contributing to the inservice professional development of teachers.

Presently, Yemeni colleges of education offer four-year preservice education programs (called concurrent programs) and sequential programs. Coursework in the concurrent programs is divided into three main components: general pedagogical preparation (approximately 41 percent of the course credits), subject-matter preparation (approximately 50 percent of the course credits), and fieldwork (approximately 9 percent of the course credits). The sequential programs require a Bachelor's degree in a subject area (e.g., physics, chemistry, or biology) followed by one or two years of education courses.

Continuing Professional Development

The unqualified status of many Yemeni teachers has prompted the MOE to organize extensive professional development programs, most of which are supported by international non-governmental education organizations. Colleges of education also provide some professional development activities. Every year, between 25 and 50 percent of teachers receive some kind of professional development (UNESCO 2011). However, these professional development activities have not been successful in improving the teaching skills of teachers and other school staff (World Bank 2010). The activities are characterized by a lack of responsiveness to the real needs of teachers because they emphasize theoretical rather than practical matters, are limited in duration, and focus on general issues instead of specific content matter and curriculum-related topics. In addition, these programs are one-off activities that train the teachers and then leave them to apply what they have learned by themselves without any professional support (UNESCO 2011; United Nations Global Education First Initiative 2013).

The need to improve teacher quality and to upgrade the knowledge and skills of teachers who do not have university degrees or who have university degrees in areas not related to education led the MOE in 2005 to develop a strategic plan for "enhancement and training," with the aim of meeting the plan's objectives by 2015 (MOE 2005). These objectives include raising the competency level of teachers with no university degrees to the level of a Bachelor's degree, raising the competency level of teacher trainers, developing the institutional capacity of training institutions and of the central office of the MOE and its regional offices, and further developing preservice and inservice institutions. The strategy includes a detailed action plan for the years 2006 to 2015 but does not include a specific focus on the needs of science teachers.

Science Teacher Professional Associations

Unfortunately, Yemen has no professional associations dedicated to supporting science education. However, there is a science education journal initiated by science educators and produced by the Science and Technology University. It provides comprehensive articles (both academic and informative) relevant to science education in Yemen and is available free in digital form from the university's website.

Issues in Upper Secondary Science Teacher Quality

Teacher Supply

Problems of teacher supply in Yemen can be attributed to the rapid developments in the education system since unification of the country in 1990, at which time access to education increased immensely, resulting in a very high demand for teachers at all educational levels (de Feiter and An'am 2007). The lack of qualified teachers is exacerbated by the problem of teacher distribution, with most unqualified teachers teaching in rural areas (United Nations Global Education First Initiative 2013). This pattern is especially problematic at the secondary level where most of the science teachers in rural schools are unqualified or are teaching out of field.

Teacher supply in Yemen presents a paradox. While many regions in Yemen suffer from a lack of teachers, close to 100,000 teachers were looking for a job in 2007 (de Feiter and An'am 2007). Reasons for this paradox include resistance to working in rural areas and a mismatch between the needs of schools and the types of teachers prepared in colleges of education. According to de Feiter and An'am, Yemen needs four types of teacher—homeroom teachers for Grades 1 to 3, "domain" teachers for Grades 4 to 9, subject teach-ers with at least one teaching subject at the upper secondary level and a subject domain at the lower secondary level, and subject teachers for upper second-ary schools. However, colleges of education prepare mainly subject-matter teachers for upper secondary schools, and these teachers refuse to teach at other levels and in rural areas. These problems have been aggravated since 2011 because of the security situation in the country, making it harder than ever for the MOE to take steps to resolve the difficulties.

Professional Education and Training Issues

The quality of science teacher preparation in colleges of education is still defi-cient because of the emphasis on theoretical rather than practical matters in science, the lack of emphasis on student-centered and inquiry-based teach-ing, and limited opportunities for teachers to experience supervised teaching before they start teaching. In addition, when these teachers start teaching, they do not have opportunities for continued professional growth because of the low quality of professional development activities available to them.

Despite the fact the MOE has developed plans and strategies to improve preservice and inservice education, the objectives of these plans have not been achieved because of implementation of the plans not materializing for a variety of reasons, including financial and security issues. Another reason for the non-realization of the strategy objectives is the low quality of students applying for colleges of education and the resistance of graduates of these colleges to teach in rural areas. These two factors, in turn, are associated with pressures on universities to accept all students eligible to attend them and to allow teachers who are MOE employees to resist orders to teach in areas in need of teachers.

Teacher quality in Yemen is still a serious problem. Most teachers do not have the required qualifications to teach science at the elementary level, while preparatory and secondary teachers may have the qualifications, but the pro-grams from which they graduate are focused on theoretical knowledge rather than on problem solving, critical thinking, and science as inquiry.

BouJaoude (2006) has shown that most science teacher preparation pro-grams in Arab countries, including Yemen, adopt an "academic/technologi-cal orientation" approach to teacher preparation. The academic orientation is "primarily concerned with the transmission of knowledge and the devel-opment of understanding" (Feiman-Nemser 1990, 221), whereas the techno-logical orientation focuses on the knowledge and skills of teaching. Yemeni teacher preparation programs are no exception. Consequently, teachers pre-pared in these programs tend to reproduce the methods of teaching they experienced rather than use reflection and problem solving. The results are an emphasis on memorization, lack of attention to using inquiry and lab-oratories, and graduating students without the scientific and creative skills necessary for living successfully in the twenty-first century. The problem of unbalanced distribution of teachers between urban and rural areas also contributes to this situation because, as mentioned earlier, most schools in rural areas are staffed by minimally qualified or unqualified science teachers (World Bank 2010).

Issues at the Chalkface

The MOE has developed curricula that emphasize a student-centered, discovery approach to science teaching, relatively high-quality textbooks and teachers' manuals, and laboratory guides for secondary-level biology, chemistry, and physics. A quick analysis of the scope and sequence of the science curriculum and the science textbooks in light of the number of periods available to teach science at all levels leads to the tentative conclusion (confirmation of which is reliant on an in-depth study of the curriculum) that coverage of topics is more important than depth of scientific knowledge. In addition, most teachers do not seem to be implementing the curriculum as envisioned by the documents. UNESCO (2011) contends that teachers tend to overuse the textbooks as sources of information rather than as resources to help students develop scientific skills and apply science in everyday life situations.

The Yemeni secondary school curriculum in general is also subject to criticism. It is viewed as theoretical and focused mainly on ensuring that students learn only the academic content they need to enter university education (World Bank 2010). Particular criticisms focus on lack of emphasis on life and career skills, problem-solving skills, and information and technology (IT) skills. Moreover, according to the MOE (2006), secondary curricula are very content-dense with not enough time to cover them in depth in the classroom. In addition, the curricula suffer from lack of proper vertical and horizontal sequencing and neglect of practical work, especially in the sciences (MOE 2008).

Al-Zanan (2014) reported that most Yemeni secondary schools do not have laboratories. Where laboratories are available, they are generally not well equipped or effectively used. According to Al-Zanan, the reasons why laboratories in schools that have them are not used include lack of a budget for consumables, high numbers of students in science classes, lack of incentives for teachers to use laboratories, small number of teaching periods allotted to science at the secondary level, emphases in curricula and examinations on theoretical and memorized scientific knowledge, and lack of science teacher preparation to use laboratories. Although the curriculum was designed to reflect a student-centered and discovery approach, this vision has not been realized because teachers do not receive training in these approaches (UNESCO 2011).

The large class sizes, outdated curricula, lack of resources, and over-reliance on textbooks and laboratory guides along with unqualified teachers have resulted in classroom environments that do not seem to support student learning. While there are no studies to gauge the performance of secondary school students on comparative international studies of science and mathematics, the results of Yemeni Grade 4 students on the Trends in International

Mathematics and Science Study (TIMSS) iterations of 2011 and 2015 were close to the lowest in the world (Education Policy Data Center 2014; MOE 2013; Zakout 2015). Unfortunately, attempts to improve the quality of classroom teaching and, consequently, the quality of student performance (see, for example, MOE 2013) have not been successful because of the aforementioned reasons and the security situation in the country.

In response to the weaknesses of the secondary curriculum, the Yemeni government developed a secondary education strategy for the years 2006 to 2015 (MOE 2006). The general strategic goal was to "provide just and equitable access to high quality and diverse secondary curriculum for graduates of basic education with the aim of reaching a graduation rate of 56% by 2015" (translated from the Arabic). The strategic goals identified were to increase access to and graduating rates from secondary education, improve the quality of secondary education in terms of student achievement, diversity of specializations, and alignment with student and career needs, and improve system governance and efficiency at all levels (school, local, and regional). The strategy has no specific goals for science education, but the second goal alludes to improving preservice and inservice teacher training programs in general.

Continuing Professional Development Issues

In addition to the problems associated with teacher preparation programs, there is evidence to suggest that inservice professional programs, though abundant, have not resulted in significant improvement in the teaching skills of science teachers. One apparent reason for this outcome is lack of an explicit theoretical and organizational framework that includes targeted professional activities and follow-up mechanisms to ensure that acquired skills are transferred to classroom practices. As a response to the dire situation of professional development, the MOE included in its 2006 to 2015 education strategies a ten-year professional development strategy, along with detailed action plans. However, as this ten-year period draws to an end, there is no evidence to suggest this strategy has resulted in positive change.

Trends and Developments in Upper Secondary Science Teacher Quality

The unavailability of progress reports makes it hard to evaluate current efforts to improve the quality of education in Yemen, although recent reports on both education and teacher quality in the country (e.g., Al-Sakkaf 2014; UNESCO 2011; World Bank 2010) indicate that neither is yet at an acceptable level. For example, at present, only 40 percent of basic education teachers hold a Bachelor's degree, there is no evidence that the quality of the colleges of education has improved, and teacher and administrator absenteeism is very high due to lack of institutional monitoring at all educational levels. In addition, textbooks and other learning materials often do not arrive until close to the end of the school year, leaving most teachers and students without them All of these factors negatively affect the quality of student learning (United Nations Global Education First Initiative 2013).

It is nonetheless evident that the Yemeni MOE has invested considerable effort in developing strategies that aim to improve the quality of the country's education system at all levels. Examples of these strategies include, but are not limited to, the Yemeni national strategy for secondary education 2006 to 2015 and the teacher enhancement and training strategy. While these strategies are detailed and well written, there is no evidence that their goals have been achieved.

There is some evidence, however, to support the claim that the problems of access to education at the pre-tertiary level in Yemen have been addressed to an acceptable level, even though certain problems remain. For example, illiteracy rates are still relatively high in the population as is scientific and technological illiteracy (United Nations Development Program/Regional Bureau for Arab States 2002, 2003; World Bank 2007). Despite problems of access having been generally addressed, education, including science education, in Yemen continues to be of a low quality across all levels of the education system.

The problem with quality of science education in Yemen is demonstrated in the large number of insufficiently qualified science teachers, the emphasis on theoretical science education and neglect of hands-on and practical activities, the nature of the science curriculum and teaching methods, lack of access to appropriate laboratory equipment and technological tools, and (most importantly) lack of adequate budgets for improving the quality of science education. These problems are particularly serious at the upper secondary stage, and especially in science, because upper secondary education is a stepping stone to either employment or higher education. Accordingly, increased access to quality education for all students and in all subjects, and especially in science, mathematics, and technology, at this stage is an important determinant not only of the quality of life that these students will experience in the future but also of the economic and cultural welfare of Yemini society as a whole.

The political and security situation is presently very unstable in Yemen. This situation has exacerbated the grim educational situation described above. However, it seems that the Yemeni MOE is continuing its efforts to remedy the situation with support from international organizations and donors. It is only when these efforts are complemented by the resolve of the Yemeni people to improve on the status quo that we can start seeing light at the end of the tunnel. After all, the future welfare of Yemen is in the hands of the young people of today. If we fail to properly care for them, the future of all of us becomes uncertain.

References

Al-Hage, M. 2006. "Teacher Education Institutions in Yemen: Current Status and Future Prospects." The seventh scientific conference on teacher preparation institutes in the Arab Nation. Cairo: Egypt, http://t-p-uqu.blogspot.com/2013/06/blog-post.html#!/2013/03/blog-post_7445.html (in Arabic).

Al-Sakkaf, Nasser. "Yemen's Teachers Aren't Making the Cut." *Yemen Times*, January 2, 2014, http://www.yementimes.com/en/1743/news/3312/Yemen%E2%80%99s-teachers-aren%E2%80%99t-making-the-cut.htm.

Al-Zanan, A. 2014. "Using Science Laboratories in Yemeni Secondary Schools: Current Status and Challenges." Master's diss, Um Al Kura University, Saudi Arabia, http://libback.uqu.edu.sa/hipres/ABS/ind15189.pdf (in Arabic).

BouJaoude, Saouma. 2006. "Bridging the Gap Between Scientists and Science Educators in the Arab Region." Report presented at the Expert Group Meeting on "Bridging the Gap Between Scientists and Science Educators," organized and sponsored by the UNESCO Office, Cairo, Egypt, from January 29–February 1, 2006.

De Feiter, L., and M. A. Anaam. 2007. "Towards the Alignment of Demand and Supply for Teachers and Teacher Education in Yemen." Unpublished paper prepared for the Dutch-Yemeni project, "The MASTERY Project" (Mathematics and Science Teacher Education Reform in Yemen). Faculty of Education, Sana'a University.

Education Policy Data Center. 2014. *Yemen National Education Profile: 2014 Update*. Washington, DC: Author, http://www.epdc.org/sites/default/files/documents/EPDC%20NEP_Yemen.pdf.

Feiman-Nemser, Sharon. 1990. "Teacher Preparation: Structural and Conceptual Alternatives." In *Handbook of Research on Teacher Education*, edited by W. Robert Houston, Martin Haberman, and John P. Sikula, 212–33. New York: Macmillan Publishing Company.

Ministry of Education and Technology. 1992. *Yemeni Education and Teaching Law*. Sana'a: Author, http://www.aub.edu.lb/main/Pages/index.aspx (in Arabic).

———. 2005. *Rehabilitation and Training Strategy*. Sana'a: Author, http://www.yemenmoe.net/TrainingStratigy.aspx (in Arabic).

———. 2006. *Yemeni National Strategy for Secondary Education 2006–2015*. Sana'a: Author, http://planipolis.iiep.unesco.org/upload/Yemen/YemenSecondarystrategy.pdf (in Arabic).

———. 2008. *The Development of Education in the Republic of Yemen*. Sana'a: Author, http://www.ibe.unesco.org/National_Reports/ICE_2008/yemen_NR08.pdf.

———. 2013. *Yemen Education Sector Plan: Mid Term Results Framework 2013–2015. A Document of Yemen Ministry of Education*. Sana'a: Author, http://www.globalpartnership.org/content/yemen-education-sector-plan-mid-term-results-framework-2013-2015.

United Nations Global Education First Initiative. 2013. *Accelerating Progress to 2015: Yemen*. Paris: The Good Planet Foundation, http://educationenvoy.org/wp-content/uploads/2013/07/YEMEN-UNSE-FINAL.pdf.

United Nations Development Programme/Arab Fund for Economic and Social Development (UNDP/RBAS). 2002. *Arab Human Development Report 2002*. New York: Author, http://www.arab-hdr.org/publications/other/ahdr/ahdr2002e.pdf.

———. 2003. *Arab Human Development Report 2003*. New York: Author, http://www.arab-hdr.org/publications/other/ahdr/ahdr2003e.pdf.

UNESCO. 2011. *World Data on Education*. Geneva, Author, http://unesdoc.unesco.org/images/0021/002117/211701e.pdf.

World Bank. 2007. *The Road Not Travelled: Education Reform in the Middle East and North Africa*. Washington, DC: Author, https://openknowledge.worldbank.org/handle/10986/6303.

———. 2010. *Republic of Yemen Education Status Report: Challenges and Opportunities*. Washington, DC: Author, http://reliefweb.int/sites/reliefweb.int/files/resources/Full_Report_4159.pdf.

Zakout, Wael. 2014. "Raising the Quality of Education in Yemen," blog, December 12, 2014, http://blogs.worldbank.org/arabvoices/raising-the-quality-of-education -in-yemen.

Further Reading

Al-Batuly, Abdulmajeed, Mohamed Al-Hawri, Martin Cicowiez, Hans Lofgren, and Mohamma Pournik. 2011. *Assessing Development Strategies to Achieve the MDGs in the Republic of Yemen.* United Nations Department for Social and Economic Affairs, http://www.un.org/en/development/desa/policy/capacity/output_studies /roa87_study_yem.pdf.

19

Bangladesh

Muhammad Salahuddin

Teaching Science in Bangladesh

Backdrop

Bangladesh is a South Asian country. Although it is only 147,547 square kilometers in size, more than 150 million people live there. Within the country, 98 percent of people identify themselves as Bengali and the rest are known as indigenous minorities. Of the 75 indigenous communities living in Bangladesh, 43 have their own language. However, Bangla is the official language of the country, and 95 percent of people use it frequently; English is considered to be the second language. Bangladesh has a per capita annual income of US$1,190. Despite a GDP real growth rate of just over 6 percent, 69 percent of people live in poverty (Ministry of Finance 2014).

Quality education is one of the key priority issues in Bangladesh (Government of the People's Republic of Bangladesh 2010). In 1974, only 27 percent of the population was literate, but this figure has since risen to 58 percent. The education system of Bangladesh is made up of five layers:

- Non-compulsory pre-primary education (one year);
- Primary education, five years' duration, free of charge, and compulsory for children once they reach six years of age;
- Junior secondary level, three years' duration;
- (Middle) secondary level, two years' duration and referred to simply as "secondary"; and
- Upper/higher secondary level (two years).

In 2018, primary education will be extended in length from five to eight years (Government of the People's Republic of Bangladesh 2010), thus making the basic education cycle an eight-year one. Primary education in Bangladesh currently has three major strands: Bangla-medium, English-medium, and madrasah education (*Ebtedayee*). There are 12 types of primary schools, including schools attached to high schools and schools attached to primary teacher training institutions. About two-thirds of primary schools are in the government sector. The Directorate of Primary Education under the Ministry of Primary and Mass Education monitors the entire primary education system. At the end of their primary schooling, students take the Primary School Certificate examination, which controls the transition to secondary schooling. About 75 percent of each primary school cohort make the transition.

At present, secondary education consists of seven years of formal schooling and encompasses three major areas (tracks): general education, madrasah education, and vocational education. The general education system is covered by Bengali-medium and English-medium schools, whereas the other two areas are covered by madrasas and vocational institutions respectively. After completing the junior secondary level, students sit a public examination called the Junior School Certificate; about 84 percent of each student cohort passes it. Three tracks emerge in Grade 9: humanities, science, and business. In most cases, teachers help students choose a track that accords with their interests and previous academic achievement.

Secondary education is designed to prepare students for progress to the higher secondary stage. The examination gateway is the Secondary School Certificate, the completion rate for which is about 75 percent. The transition rate to higher secondary schooling is currently a little in excess of 50 percent (Bangladesh Bureau of Educational Information and Statistics 2013). Higher secondary education is offered by so-called intermediate colleges and by the intermediate section of colleges accredited to award Bachelor's degrees. Students enter one of three tracks (humanities, science, and business), each of which prepares students for tertiary-level education. At the end of their higher secondary schooling, students sit the Higher Secondary Certificate examination.

School Science

During Bangladesh's colonial period (1757–1947), science education was absent from the school curriculum. However, in 1961, science education (physics, chemistry, and biology) was introduced at the secondary level, and throughout the 1960s, when Bangladesh was part of Pakistan, ongoing effort was made to include science content that would prepare students to secure good jobs and take part in higher education in Western countries. Bangladesh became an independent nation in 1971. The first post-independence education commission, established in 1972, focused on improving education. The commission's report, published in 1974, emphasized the need for science education. It stated the main aim of science as being "to expedite the overall development of society, and not merely to make a particular group of men powerful through application of knowledge derived from the unveiling of the secrets of nature" (Bangladesh Education Commission 1974, 105). The authors of the report recommended that science education should be mandatory for Grades 3 to 5 primary students and for all secondary students. They also recommended updating science syllabi, introducing new methods in science teaching, and ensuring an adequate supply of scientific instruments and books to all educational institutions.

Unfortunately, these recommendations were not implemented due to political instability. Five more policies were published in 1978, 1988, 1997, 2000, and 2003 respectively. Science education was again highly emphasized in each, but only a few of the reports' recommendations were properly implemented. More recently, and in line with Bangladesh's attempts to develop a comprehensive education policy, the country's National Education Policy 2010 (Government of the People's Republic of Bangladesh 2010) stated that the objective of science education in Bangladesh should be to prepare learners to pursue knowledge and creativity in a way that enables them to reach international standards of achievement in science.

It can certainly be said that science education is seen as an important part of schooling in Bangladesh. It is a compulsory subject for primary students and for students at the junior secondary level. At the secondary level, only students in the science track study physics, chemistry, and biology as separate subjects, while humanities and business studies students study the subject "general science." At the higher secondary level, only students in the science track study science subjects. These include physics I and II, chemistry I and II, biology I and II, and higher mathematics I and II.

Assessment for the science subjects (physics, chemistry, and biology/ higher mathematics) has two major components at the secondary and higher secondary levels: a written examination and a practical examination. Physics students, for instance, can gain 75 marks of the maximum attainable score of 100 through written questions (40 for open-ended questions and 35 for multiple-choice items) and 25 marks through the practical test. The examination is administrated by the secondary and higher secondary education boards.

Teacher Academic and Professional Education and Training

The Public Service Commission (PSC) recruits higher secondary teachers for government schools while the Non-Government Teacher Registration and Certification Authority is responsible for recruiting and deploying teachers in non-government secondary schools. Entry-level requirements for persons wanting to teach at the secondary and higher secondary levels are a minimum of a Bachelor's degree or equivalent for junior secondary and secondary education and a Master's degree for higher secondary education. Candidates holding a Bachelor of Education (BEd) with or without a Master of Education (MEd) alongside a subject degree get preference. Candidates must present degrees in which they have achieved at least a third-class pass.

Although there is no preservice training for people wanting to teach at any level of the education system in Bangladesh, teachers are required to receive inservice training. Bangladesh currently has 209 training institutions. These include the following:

- Two institutes of education and research under the public universities;
- Fifty-four primary training institutes;
- One hundred and eighteen teacher training colleges;
- One technical teacher training college;
- One vocational teacher training college;
- Twenty-nine physical education colleges;
- Five higher secondary teacher training institutes; and
- One madrasah teacher training institute.

Their offerings include the Certificate in Education, Diploma in Education, Bachelor of Education, Master of Education, and Bachelor of Physical Education. They also provide subject-based training, ICT-related training, and training in assessment.

The National Academy for Educational Management (NAEM) plays a major role in providing basic training and subject-based training for newly recruited teachers at the secondary and higher secondary levels. NAEM provides foundation training (120 days' duration), an educational research methodology course (30 days) for government college teachers, a senior staff course on educational management (45 days), advanced courses on education and management (45 days), secretarial work and office management courses (14 days), and educational planning and development courses (30 days) for assistant and associate professors of government colleges. NAEM also provides educational administration and management courses, a Communicative English course for principals of upper secondary and secondary-level institutions, and training courses on ICT for college lecturers and Ministry of Education officers. Each of these courses is three weeks in duration. NAEM furthermore conducts professional leadership training (21 days) for heads of upper secondary and secondary-level institutions. It carries out this work under the supervision of the Teaching Quality Improvement in Secondary Education Project (TQI-SEP 2007).

Continuing Professional Development

TQI-SEP plays a key role in the provision of continuing professional development (CPD) for teachers working at the secondary and upper secondary levels. The aim of the project is to further develop secondary education teachers' subject-based knowledge and skills (TQI-SEP 2007). In pursuit of this aim, TQI-SEP has formed collaborations with the University of Dhaka's institute of education and research, 14 of Bangladesh's teacher training colleges, the country's five higher secondary teacher training institutes, and its madrasah teacher training institute. It also works closely with NAEM. During the first phase of the project (2007–2011), the University of Dhaka's institute of education and research provided training known as "training of trainers" to "master trainers" at the teacher training colleges, higher secondary teacher training institutes, madrasah teacher training institute, and NAEM. The master trainers of all these organizations except NAEM then provided CPD for secondary teachers. During the current second phase of the project (2012–2017), the master trainers are providing CPD for teachers, including biology and mathematics teachers, at the upper secondary level. Various experts and consultants have been preparing a CPD training manual for use at the upper secondary level, which at the time of writing was expected to be implemented during 2015.

The subjects for which TQI-SEP provides CPD training at the secondary level include Bangla, English, social science, physics, chemistry, biology, mathematics, and agriculture. Training encompasses 14 days (24 sessions), and the manual includes 32 lesson plans for each subject. Training seeks to instill basic understanding about basic issues such as the teaching profession, lesson planning, teaching aids, the teaching–learning process, classroom management, participatory and inclusive teaching, gender issues, classroom questioning, and techniques of assessment. After successfully completing CPD-I (subject based), teachers can take CPD-II training directed toward further improvement in the above listed areas. This training is often needed because a good number of teachers have difficulty implementing CPD-I at classroom level. TQI-SEP accordingly develops its CPD-II content with reference to CPD-I feedback and question-and-answer sessions about CPD-I. It also offers information about teaching methods and aids and provides guidance on formative assessment. Effort is made throughout these courses to enhance the quality of teaching and learning by nurturing positive attitudes and practices at the classroom level.

The CPD courses at the secondary level take place at the teacher training colleges, the higher secondary teacher training institutes, and the madrasah teacher training institute. During the year that ran from November 2013 to November 2014, more than 2,000 biology teachers and 2,700 mathematics teachers received CPD training. Four hundred and seventy-one of the biology teachers and 237 of the mathematics teachers were female.

The government has the right to take action against teachers at the upper secondary level who infringe their conditions of employment. These infringements include examination-related issues, inferior educational performance, and absenteeism. If government or government-supported colleges contravene government standards, the government can stop their financial assistance. However, it is rare in Bangladesh for any kind of official action to be taken against college teachers.

Science Teacher Professional Associations

Within the local context, a number of professional associations support teachers' comprehensive development. These are the Bangladesh Civil Service

General Education Association, Bangladesh College Teachers' Association, Bangladesh College Teachers' Association Federation, Bangladesh Teacher Association, Bangladesh College University Teachers' Association, Bangladesh Technical College Teachers' Association, and Bangladesh Jomiaytul Modarresin. Bangladesh has no science teacher associations per se.

All of these organizations have political affiliations and negotiate teachers' conditions of employment with the government. Consequently, their focus is on teachers' wages, opportunities, and facilities rather than on teachers' professional development. The Bangladesh Teacher Association, one of the largest teachers' associations in the country, works for the welfare of teachers, students, and schools. It takes action on behalf of teachers against discrimination and harassment and for teachers' legal rights. It also helps teachers by giving them appropriate advice and guidance for their retirement; its overarching aim is to facilitate quality education through quality teachers. As part of its commitment to developing the education sector, the association has recommended decentralizing the education sector, initiating radical changes in the existing examination system, and establishing a teachers' recruitment commission and a permanent education commission.

Issues in Upper Secondary Science Teacher Quality

Teacher Supply

The main supply of teachers for the secondary level is from the universities, of which there are 86. As mentioned earlier, the Public Service Commission recruits teachers for government upper secondary schools and colleges while the Non-Government Teacher Registration and Certification Authority mainly recruits teachers for non-government secondary schools and colleges. Candidates must have a subject-related degree, the minimum being either a second-class (or equivalent cumulative grade-point average) three-year Bachelor's (Pass) degree, or a second-class (or equivalent) four-year Bachelor's (Honors) in a relevant subject (physics/chemistry/biology/mathematics) from any recognized university. Teachers for government secondary schools and colleges are recruited through the Public Service Commission examination, which involves a multiple-choice test in the first phase, a written test (general and subject related) in the second phase, a viva examination in the third phase, and a medical test in the final phase. Candidates should be between the ages of 21 and 30 years at the time of application and must be citizens of Bangladesh. Teachers for non-government secondary schools and colleges must complete the subject-related multiple-choice test and written test administered by the Non-Government Teacher Registration and Certification Authority. They must also undertake a viva examination at a school. Around half (49 percent) of secondary-level science teachers have Bachelor's (Pass) degrees while 43 percent have a Master's degree in their respective science subjects (Tapan, Rahman, and Ahmed 2014).

Upper secondary science teachers receive a monthly salary, house rent, medical allowance, and festival bonuses. Government college teachers are also entitled to retirement benefits (superannuation). However, most teachers at the upper secondary level are not satisfied with their job, the main sources of dissatisfaction being salary, social status, and workload. Most teachers receive less than US$150 a month, and many teachers' families experience financial problems. However, teachers have a relatively high social status because teaching is considered a respectable profession in Bangladesh. There are no national statistics on teacher supply, but there is no evidence of a critical shortage. However, there is a lack of qualified teachers in science, particularly in rural areas (Tapan 2010).

Professional Education and Training Issues

Although, as noted earlier, Bangladesh does not have a system of preservice teacher education and training, professional education for science teachers is gradually developing in the country. A good number of the government and private teacher training institutions provide inservice teacher training for all subject teachers, including science teachers (Bangladesh Bureau of Educational Information and Statistics 2013). In addition, secondary education and science development centers have been established within the campuses of nine teacher training colleges. Their purpose is to provide inservice training to secondary and upper secondary teachers.

About 10,000 teachers every year undertake inservice training of two to three weeks' duration in physical and biological sciences, mathematics, social science, and Bengali (Tapan 2010). Some private organizations offer short-term science teacher training in various subjects. The University of Dhaka offers secondary and upper secondary school teaching professionals a Master of Education evening course on science, mathematics, and technology in education through its institute for education and research.

Secondary school teachers nonetheless require more training in science content, practical work, teaching methods, and assessment (Rashid 2001). Upper secondary science teachers need an even greater amount of training in science education and teaching (Tapan 2010), but they do not currently have enough opportunity to pursue it. However, programs designed to address this situation should begin appearing in 2015.

Issues at the Chalkface

Most upper secondary science teachers in Bangladesh use the lecture method of teaching. They sometimes do use more participatory approaches and set peer-based and group-based work, but they rarely have sufficient motivation to conduct practical work. Science laboratories and equipment bases are inadequate throughout the country, a situation that has contributed to the rapid decline over the past few years in the number of students taking science subjects at secondary, upper secondary, and higher education levels (Ashraf 2008). The lack of facilities such as laboratories and equipment, as well as of qualified teachers, is more acute in rural areas than in urban areas (Tapan 2010). A study conducted through the Foundation for Research in Educational Planning and Development (Tapan et al. 2014) found that while most schools had science laboratories, specialized laboratories for the individual sciences were much less prevalent. Only 41 percent of schools had physics labs, and only 37 percent had separate laboratories for chemistry and biology. Necessary materials and equipment were found in only 53 percent of schools, a deficit that was impeding use of the laboratories. Seventy-three percent of schools were using their laboratories no more than one day each week; others no more than three times a week. Only 8 percent of the schools were using their laboratories on a regular basis.

The study also found that only 67 percent of schools were regularly using their laboratories for practical lessons, and that only a few schools (10 percent) could avail themselves of lab demonstrators; in the majority of schools (76 percent), science teachers themselves were conducting practical classes. Eight percent of rural schools had no laboratory facilities at the secondary and upper secondary levels. Practical classes were not being conducted properly due to lack of facilities and teacher motivation. Despite these shortcomings, another study (Ashraf 2008) found that 80 percent of students studying science received pass marks simply for attending the practical examination, even though the schools were supposedly conducting those examinations in accordance with ministerial guidelines.

In 2012, Dr. Satyabratra Roy, having received financial assistance from the Foundation for Research in Educational Planning and Development, evaluated the status of practical science education at the secondary level of Bangladeshi schools (Roy 2014). He found that 31 percent of schools and 50 percent of madrasas had no laboratory while 37 percent of urban and only 7 percent of rural schools had subject-specialized laboratories. Half of the laboratories had only one door (two doors being required), and 73 percent of the laboratories did not present viable learning environments with respect to features such as a display board, graph board, gas and water supply, water basin, water tap, gas tap, smoke detectors, tool box, first aid box, and fire extinguisher. Roy also reported that 34 percent of schools and 20 percent of madrasas did not hold physics practical classes; the corresponding figures for chemistry were 40 percent and 53 percent and for biology were 60 percent and 67 percent. The majority of secondary and upper secondary schools had no lab assistants.

The school principals who participated in Roy's survey said these deficits were just two of the many problems adversely affecting science teaching and learning in their schools. The additional problems included lack of technical and chemical materials, lack of skilled science teachers, practical classes not being scheduled into the class routine, pressure on teachers brought about by large class sizes and heavy timetables, lack of teacher training in conducting practical classes, low teacher motivation, inaccurate assessment of student achievement, financial corruption (i.e., teachers being bribed to give passing grades), and lack of an overall adequate budget. Finally, more than half of the schools and two-thirds of the madrasas in Roy's study had no practical assessments during semester examinations. Also, in most cases where schools did have practical classes, these tended to be only arranged, and hurriedly so, shortly before the final examination (see also in this regard Ashraf 2008).

Continuing Professional Development Issues

As stated above, teachers obtain basic training after joining the profession. The Ministry of Education intends to introduce continuing professional development for all teachers at the upper secondary level during 2015. At present, the ministry is developing training manuals for physics, chemistry, biology, and mathematics. The content of these manuals lists and describes teachers' work and responsibilities, discusses the teaching–learning process, emphasizes the importance of lesson planning and evaluation, and focuses on classroom management. The manuals also cover teaching aids and their use in the classroom and how to develop such aids cheaply using local materials. The manuals furthermore highlight the development of yearly plans for specific subjects and provide advice on how to ensure inclusive teaching and learning-friendly environments in classrooms.

The upper secondary science manual additionally accentuates gender issues with respect to developing and ensuring participatory learning environments in the classroom. Assessment and feedback-related issues, open-ended questions, written and practical examinations, interconnected topics, and other issues are also part of the manual's content. The manual is being developed in line with Bangladesh's National Education Policy of 2010 and the country's new curricula program. University teachers, subject experts, educationists, training experts, and TQI-SEP officials are all involved in the process.

Teachers participating in the training set out in this manual will be required to set aside 24 days to complete the course. At the time of writing, training was expected to begin during the first quarter of 2015. The agencies responsible for providing this training (all working under the supervision of TQI-SEP II) include the aforementioned 14 teacher training colleges, 5 higher secondary teacher training institutes, and the madrasah teacher training institute.

Trends and Developments in Upper Secondary Science Teacher Quality

Upper secondary science education in Bangladesh has developed gradually over the last five decades. In 1961, science education was accorded a strong emphasis in the school curriculum, a development that prompted the addition of physics, chemistry, and biology as different subjects, each with theoretical and practical components. Major reforms occurred in 1983 and 1995. Today, upper secondary science students can study physics, chemistry, biology, or higher mathematics.

Since the beginning of the twenty-first century, the government has undertaken many initiatives directed toward developing science education in Bangladesh. The new education policy developed in 2010 emphasizes the modern world and competitive global market (Ahmed and Salahuddin 2014). In 2012, a new curriculum accompanied by new textbooks was introduced into upper secondary schools. Changes have also been made to the country's system of assessing educational achievement, with students from junior secondary level to upper secondary level now being assessed through questions requiring open-ended responses.

The availability of quality teachers has been one of the main obstacles facing science education in Bangladesh since independence (Ashraf 2008; Rashid 2001; Roy 2014; Tapan 2010). Tapan (2010) attributed the lack of effective change in science education at the secondary level to this problem. In every education policy, teacher training has been assigned high importance, but the reality has not reflected this. Until very recently, secondary teachers, including science teachers, have had no continuing professional development, but major effort to ameliorate this situation was implemented in the first quarter of 2015. Today, various training programs in teaching and learning are being conducted all around the country through the teacher training colleges, the higher secondary teacher training institutes, and the madrasah teacher training institute.

At present, almost all science teachers for the upper secondary level are recruited through a recruitment examination, and all candidates must have at least a BSc. These individuals should be well disposed to gain benefit from inservice training on different issues and to make use of the new textbooks and teacher guides developed in pursuance of quality education. Practical classes and examinations for science subjects are gradually being given more attention, and the government is providing more funding than it has in the past for establishing upper secondary science laboratories. The convergence of these various interventions bodes well for science teacher quality in Bangladesh in the years to come.

References

Ahmed, Shah Shamin, and Muhammad Salahuddin. 2014. "Impact of Globalization on Education Policy: Bangladesh Perspective." *The Dhaka University Studies* 68 (2): 117–28.

Ashraf, Shamim. 2008. "State of Science Education in Bangladesh: Current Status and Future Trends." *The Daily Star*, July 27.

Bangladesh Bureau of Educational Information and Statistics. 2013. *Bangladesh Educational Statistics 2012*. Dhaka: Ministry of Education.

Bangladesh Education Commission. 1974. *Bangladesh Education Commission Report 1974*. Dhaka: Ministry of Education, http://www.banbeis.gov.bd/devnetsolutions /BANBEIS_PDF/BANGLADESH%20EDUCATION%20COMMISSION%20 REPORT %20%20-1974.pdf.

Government of the People's Republic of Bangladesh. 2010. *National Education Policy 2010*. Dhaka: Ministry of Education, http://planipolis.iiep.unesco.org/upload /Bangladesh/Bangladesh_National_Education_Policy_2010.pdf.

Ministry of Finance. 2014. *Bangladesh Economic Review*. Dhaka: Author.

Rashid, R. 2001. "Determination of Training Needs of Physics Teachers at the Secondary Level." Master's diss., Institute of Education and Research, University of Dhaka.

Roy, Satyabratra. 2014. *Evaluating the Present Status of Practical Science Education in Bangladesh at Secondary Level*. Dhaka: Foundation for Research in Educational Planning and Development (FREPD).

Tapan, D. M., D. H. Rahman, and Shah Shamin Ahmed. 2014. "Problems of Science Education at the Secondary Level in Bangladesh" (unpublished manuscript). Dhaka: Foundation for Research in Educational Planning and Development (FREPD).

Tapan, M. Shahjahan Mian. 2010. "Science Education in Bangladesh." In *Handbook of Research in Science Education in Asia*, vol. 4, edited by Yew Jin Lee, 17–34. Rotterdam: Sense Publishers.

Teaching Quality Improvement in Secondary Education Project (TQI-SEP). 2007. *Subject Bases Training of Class Teachers: Secondary Physical Science Training Manual*. Dhaka: Ministry of Education.

Further Reading

Choudhury, S. K. 2009. "Problems and Prospects of Science Education in Bangladesh." In *AIP Conference Proceedings* 83–84. College Park, MD: AIP Publishing.

Tapan, M. Shahjahan Mian. 2010. "Science Education in Bangladesh." *Handbook of Research in Science Education Research in Asia*, vol. 4, 17–34, edited by Y. J. Lee. Rotterdam: Sense Publishers, https://www.sensepublishers.com/media/1167-the -world-of-science-education.pdf.

20

Cambodia

Chan Oeurn Chey and Sitha Chhinh

Teaching Science in Cambodia

Backdrop

The Cambodian education system has a fragmented history due to a succession of internal and external disruptions, although its development during the Angkor era (ninth to twelfth centuries) outperformed the development of other countries in the region. During the colonial period up to 1953 and thereafter until 1975, the formal education system was modeled closely on the French system, and the country enjoyed a high quality of education, albeit with limited access (Ayers 2000).

The most devastating circumstances were experienced during the Khmer Rouge regime from 1975 to 1979. During this period, the physical infrastructure, organizational system, and human resource base were targeted for destruction in an attempt to build a utopian society that the Khmer Rouge called Angkar. School buildings were either destroyed or used for non-educational purposes such as communal centers for collective eating. Teachers, scholars, and professionals were treated as enemies of the Angkar and executed or killed through hard work or starvation (Extraordinary Chambers in the Court of Cambodia 2010). Educational materials such as books and documents were destroyed. According to the records of the Ministry of Education, Youth, and Sports (MoEYS), by the time the regime was toppled in early 1979, 90 percent of schools had been demolished and 75 percent of teachers, 96 percent of tertiary students, and 67 percent of all elementary and secondary students had died (cited in UNESCO 2008a). The destruction inflicted by the Pol Pot regime had an extremely negative impact on the development of Cambodia in general and on education in particular.

Reconstruction of the Cambodian education system started after the fall of the Khmer Rouge in 1979. During the 1980s, a general education system was adopted. It consisted of four years of primary education, three years of lower secondary education, and three years of upper secondary education. The system was reformed in 1986 with the adoption of five years of primary education. This structure remained in place until 1996.

Currently, Cambodian formal education consists of preschool education (three years), which is non-compulsory, primary education (six years), and secondary education (three years for lower and three years for upper secondary). The contribution of private schools to general educational expansion is evident in Cambodia, where the private sector has expanded progressively during the last decade. Generally, preschool education provides for children from three to five years of age, and primary education caters for children from age six. After completing the basic education cycle (six years of primary and three years of lower secondary schooling), students can continue to upper secondary education or secondary-level vocational training programs.

Due to a shortage of school buildings, lower secondary school buildings may share the same campus with a primary school or be located independently. In some cases, lower secondary and upper secondary schools are integrated into one entity, named simply high school. The enrolment ratios for primary, lower secondary, and upper secondary are 123 percent, 54 percent, and 27 percent respectively. The completion rates for the three levels are 87 percent, 41 percent, and 27 percent, respectively (MoEYS 2013).

There is no transition examination from the primary to the lower secondary level. However, students take a national examination at the end of Grade 9. In 2014, the examination was conducted by schools themselves with nominal supervision by officials from provincial offices of education. Students who pass the examination receive a certificate signifying completion of lower secondary school. The certificate also serves as the qualification determining entry to upper secondary school.

National assessment (NA) in Cambodia provides a considerable amount of information about learning outcomes and allows comparisons with student achievement in other countries (Ho 2013). Cambodia is developing its NA systems in line with knowledge gained from international experience and work with international partner agencies. In 2013, Cambodia carried out national standardized assessments of students' learning achievement in Grades 3, 6, and 8 Khmer language and mathematics. Student assessment is accorded high priority in the MoEYS's Education Strategic Plan 2014–2018 (MoEYS 2014). The plan pays particular attention to strengthening regular classroom and national assessment tests and to reforming national examinations. The MoEYS is also committed to participating in international assessments of educational achievement, such as the South East Asian Primary Learning Matric (SEA-PLM) and the OECD's Programme for International Student Assessment (PISA). The results of these studies will be used to inform educational development in Cambodia.

Cambodia's higher education sector caters for high school graduates who want to continue their studies. Universities offer two-year Associate degrees, four-year Bachelor's degrees, and seven-year Medical Science degrees.

Technical and vocational institutions offer courses for both high school completers and non-completers. Technical and vocational education and training programs offer different levels of training in various areas, including motor mechanics, computer technology, agricultural technology, electricity, electronics, and civil engineering (UNESCO 2008b).

The education law adopted in 2007 incorporated non-formal and informal education into the education system. The non-formal education department of the MoEYS plays an important role in providing literacy and life-skill programs as well as short-term vocational training programs throughout the country in cooperation with both local and international organizations. These programs were introduced in response to the high number of students dropping out of schooling after they had completed their primary education (MoEYS 2013).

Cambodia has entered a period of "demographic dividend" given the many potentially productive young people in its population. The main priority for the Cambodian government is to provide productivity-enhancing skills and also life skills to young people so that they can contribute to social and economic development, especially by attracting investment in the high value-added sectors such as electronics and agro-business industries (MoEYS 2014).

School Science

According to the policy for curriculum development 2005–2009 (MoEYS 2004), the school curriculum aims to equip students with skills conducive to balanced intellectual, spiritual, mental, and physical growth and development. Through science education, students should come to appreciate the value and importance of science, technology, innovation, and creativity and be able to protect and preserve their natural, social, and cultural environments (MoEYS 2004). In order to achieve these goals, the school curriculum has been developed to enable students to attain a high level of technical and scientific knowledge and skills in the various science subjects (physics, chemistry, biology, and Earth and environmental studies). The aim of science education at all school levels is for students to acquire not only a basic understanding of the natural world and scientific principles but also associated everyday life skills.

Science subjects are taught across the entire K–12 spectrum. However, at the primary level, science is taught as an integrated subject by the homeroom teacher; from Grade 7 on, science is taught as an independent subject by specialist teachers. In Grades 11 and 12, students are divided into two streams—a science stream and a social science stream. The science stream is the more popular among students.

In Grades 7 to 9, students develop their general knowledge base in science through topics such as how plants reproduce, grow, and obtain nutrients (biology), the classification and structure of matter (chemistry), the properties of light, sound, and different forces (physics), and the structure of Earth and environmental issues of national importance (Earth and environment science). In Grade 10, students study Khmer, mathematics, science, social studies, foreign languages, and physical and health education. They also participate in a sport and life skills program that includes art education. In Grades 11 and 12, students must take four compulsory subjects, including Khmer literature, basic or advanced mathematics, English or French, and physical and health education and sport. Those who select basic mathematics must select four other elective subjects, while those who select advanced mathematics must select three other elective subjects. Electives include the sciences, social science subjects, and vocational-technical subjects (MoEYS 2004).

Cambodia revised its science curricula in 2005/2006 and rewrote the corresponding textbooks. The country also introduced textbooks for Grades 10, 11, and 12 in 2008, 2009, and 2010 respectively. The revised texts incorporated new topics and placed particular emphasis on higher cognitive learning in the science context. The time allocated for science subjects was increased from two to three hours a week in Grade 10, and from three to four hours a week in Grades 11 and 12.

Secondary education resource centers (SRSs) have been established in each province since 2011 in order to provide both teachers and students with science education services. Each SRS serves a network of up to eight secondary schools, and each has two science laboratories equipped with modern equipment and materials for teachers and students to carry out experiments. Each science laboratory has one technical assistant. All activities are guided by a teacher. The two computer labs in each center provide students and teachers with access to multimedia materials and programs designed to enhance teaching and learning.

Currently, the quality of science education at secondary levels is low as judged by the most recent and reliable source of information—the national examination for Grade 12 in 2014; less than 10 percent of the students managed to achieve a pass in science. This poor outcome for science education is due to many reasons, such as a shortage of teachers and teaching materials, poor pedagogical practices, and low student motivation. This poor performance is evident in both the teaching and learning of science. Many students who subsequently enroll in science and science-related subjects at the higher levels have limited knowledge in science, which in turn makes it difficult for the universities to provide a high quality of education.

Teacher Academic and Professional Education and Training

Individuals wanting to become a government school teacher at any level must undertake a teacher training program at a teacher training center. Cambodia has 1 teacher training center for preschool teaching, 18 provincial teacher training centers for primary school teaching, 6 regional teacher training centers for lower secondary teaching, and the National Institute of Education for upper secondary school teaching. To enter a teacher training program, candidates must have, as a minimum, 9 years of basic education; however, most policy documents now stipulate 12 years of general education. A study by the World Bank in 2014 showed that of the 1,967 trainees at the provincial teacher training centers, 784 (40 percent) attended the "nine plus two years" training program (World Bank 2014). Candidates for lower secondary teaching must hold a certificate of general secondary education. NIE takes only candidates with a Bachelor's degree in any subject, but candidates must also pass the entrance examination.

The NIE, which changed its name in 2004 from Faculty of Pedagogy of the Royal University of Phnom Penh, provides one-year teacher training programs in 18 specialist subjects, including science subjects. Trainees are required to undertake two related teaching majors, such as chemistry and biology or physics and mathematics. Once trainees are deployed as teachers, they can teach these two subjects as their first and second majors. In 2014, the MoEYS decided to delete the second subject due to most high school teachers not teaching that subject. However, teachers in rural schools often have to teach both their subjects due to teacher shortage (Benveniste, Marshall, and Araujo 2008).

In addition to studying their major subjects, trainees are required to undertake common courses such as library management, pedagogy, classroom management, information and communication technology, and languages. Trainers are assigned to a designated upper secondary school for a one-month practicum on two occasions. During the training program, all trainees have to undertake basic skills training, professional skills training, teaching methodology, teaching practice, and pedagogical research. Within the upper secondary teacher training program, these components form respectively 33 percent, 28 percent, 14 percent, 23 percent, and 2 percent of the total program time allocation (MoEYS, 2010). Not surprisingly, subject content studies are the major focus of training in the centers. Once candidates pass the entrance examination that allows them to train at the centers, they become government employees for the rest of their lives. The centers require trainees to sit a graduation examination, but this serves the purpose of deployment policy rather than a determination of training results. The highest-scoring candidates are given priority to select the schools they most want to teach at.

Continuing Professional Development

Once in the teaching profession, teachers have almost no systematic and recognized opportunities for ongoing professional growth. Although an institutionalized inservice training and education (INSET) arrangement has been discussed for many years, the only INSET currently available is an onsite, irregularly offered short-term program with limited effectiveness; long-term, systematic, and regular INSET has yet to be implemented (MoEYS 2015).

Teachers who trained under Cambodia's previously offered "twelve plus two" training program can upgrade their teaching qualification through a higher education institution. This program leads to a Bachelor's degree in education. As yet, no systematic record of the number of teachers who have gone through this form of professional development has been kept. Private higher education institutions in Cambodia offer flexible bridging course arrangements to a degree program. When studying toward a degree, teachers have to bear all costs, including tuition fees, transport, and accommodation for a period of two and a half years. Teachers who are academically successful during training for lower secondary teaching can attend a one-year course at the NIE. The course allows them to upgrade their qualification to one that allows them to teach in upper secondary schools.

Because of the lack of graduate programs in science subjects, upper secondary school science teachers who wish to gain graduate degrees can only do so if they change their major from subject specialist to administration or management. The only institution in Cambodia that offers graduate programs in science subjects is the Royal University of Phnom Penh. However, the university's medium of instruction is English—a barrier for most high school teachers.

Science Teacher Professional Associations

In Cambodia, teacher professional associations are effectively nonexistent. However, the secondary education resource centers provide science teachers with a good environment in which to share their knowledge and teaching experiences.

Issues in Upper Secondary Science Teacher Quality

Teacher Supply

During its efforts to restore Cambodia's education system, which included recruiting and training teachers as quickly as possible, the Cambodian government was forced to adopt several teacher training programs that initially consisted of just a few months. This method gave way to a more systematic program consisting of three years of formal schooling and one year of training during the difficult time of the 1980s. This approach gave way in turn to the current practice of "nine plus two" and "twelve plus two" for basic education teachers and a Bachelor's degree plus one year training for upper secondary school teachers. As a result of the quick rebuilding of the system, the education service is served by a teaching force whose qualification levels are much lower than those deemed appropriate in the country's teacher policy action plan (MoEYS 2015). The plan stipulates at least a Bachelor's degree for basic education teachers and a Master's degree for upper secondary school teachers.

Cambodia's entire education system currently has 89,000 teachers, of whom 31 percent are lower secondary teachers and 13 percent are upper secondary teachers. Only 27 percent of secondary teachers possess a degree, and only 56 percent of secondary teachers completed upper secondary education themselves (Madhur 2014). These percentages raise questions about many teachers' ability to acquire advanced knowledge and skills. More than 2,000 educational personnel move out of teaching every year due to many reasons, retirement among them. These factors have led to the implementation of double-shift workloads for teachers (Choun 2014).

Teacher motivation is also an issue; teachers earn around 40 percent less than comparable professionals (World Bank 2014). Many teachers have to work at an additional job in order to support themselves and their families. Recruitment is a major problem due to the low salary, and there is no policy in place to encourage and support gifted science students to undertake a career in science teaching (or indeed in scientific research). Most talented science students accordingly select other professions for a better future. Most trainees entering teacher training centers for basic education teacher training have poor Grade 12 transcripts (D or E grades). However, the trainees who enter NIE have to pass the entrance examination.

Due to the uneven development of the country in favor of the urban areas, teacher deployment throughout Cambodia presents the MoEYS with a serious problem (World Bank 2014). Teachers tend to move back to their district of birth or larger provincial urban schools after a few years of being deployed to their first school of service (Benveniste et al. 2008). Teacher real-location from under-staffed and under-resourced schools to better staffed and resourced schools leaves schools in rural areas with a serious shortage of teachers, especially in the upper secondary schools. Students in rural upper secondary schools are therefore often taught by teachers qualified to teach only in lower secondary schools. The use of teachers with lower secondary credentials to teach upper secondary students has created many teaching as well as learning problems. The seriousness of the problem becomes even more evident when it is realized that the school leavers who enter regional teacher training centers to train as lower secondary school teachers are those whose high school examination results are in the lowest percentile ranks (Chhit 2013; World Bank 2014).

Professional Education and Training Issues

Once new teachers have been deployed to a school, their NIE trainers in theory have to monitor their performance for a period of one year. In practice, however, beginning teachers do not receive any support through an induction program (Chealy et al. 2014; World Bank 2014). Furthermore, teachers are inspected or assessed only on matters related to classroom routines and school administration. The shortage of trained teachers continues, and teachers receive little if any training in the newly developed school curriculum. There is a lack of appropriate preservice training in pedagogical skills and teaching methodologies, particularly those promoting student-centered learning environments (Madhur 2014).

Issues at the Chalkface

If Cambodia is to realize high-quality science education, it needs to have good classroom environments, appropriate physical facilities and science teaching methodology, and an adequate supply of teachers. During the 2012/2013 academic year, the number of upper secondary schools increased to 422. Student-to-teacher ratios between rural areas and urban areas continue to differ. The current average ratio in lower secondary schools across Cambodia is 20 students per teacher; the corresponding ratios in rural and urban areas are 22 and 17 respectively. Although the student-to-teacher ratio is lower in upper secondary schools (Madhur 2014), it is as high as 60 students per teacher in some areas due to the limited number of classrooms available and/ or large numbers of students enrolled in the schools.

The national curriculum in Cambodia has been revised and improved regularly. The main objective of this development is to ensure that students leaving school possess adequate technical skills and soft skills. Goals in regard to science are that students should appreciate the value and importance of science, technology, innovation, and creativity and are equipped with high levels of knowledge and skills in physics, chemistry, biology, and Earth and environmental science (MoEYS 2004).

As part of its efforts to achieve these goals, the MoEYS produced two series of books— student books with upgraded content and teachers' editions with the same content as the student book but also containing clear explanations of the learning objectives for each chapter, the time needed to cover the content in each chapter, and instructions for teaching. As a result, teaching styles have shifted slightly toward encouraging more thinking and reasoning skills. However, the experimental-based and scientific inquiry-based learning processes such as observing, inferring, classifying, predicting, measuring, questioning, interpreting, and analyzing remain underdeveloped owing to lack of teaching facilities, lab equipment and lab instructions, large numbers of students in a class, and time availability.

As mentioned earlier, science curricula were revised in 2005/2006 and textbooks were rewritten. However, the content and the format of the science textbooks and the methodology of teaching and learning science presented in the teacher books focus primarily on rote learning of scientific facts as a foundation for further studies. Due to their limited knowledge of subject content, most teachers adopt a formulaic approach to practical exercises and problem solving (Japan International Cooperation Agency 2012). It is traditionally observed that a popular teacher in a school is one who has committed subject content and formulae to memory.

Some studies indicate that there are teachers who do not understand the content of the textbooks to a level that allows them to confidently explain that content to their students and that this situation is especially applicable to science and mathematics (Japan International Cooperation Agency 2012). Such teachers typically skip lessons or pages. The problem is partly attributable to the content and structure of the textbooks themselves, but many teachers and students appear to lack a clear understanding about the purpose of science education and natural science in the curriculum. Analytical reports reviewing the relevance of Cambodia's science curricula to daily life and connectivity with further studies have identified deficiencies in the content and pedagogy of those curricula (see, for example, Chealy et al. 2014; Hout 2013). Numerous topics are deemed as being neither relevant to daily life nor leading to higher education studies. Content, moreover, is structured in a way that tends to encourage rote learning of facts despite the MoEYS policy stating that the curriculum should emphasize active and applied learning for deep knowledge acquisition and application.

Another criticism of school science relates to skill development. Using Bloom's Taxonomy as a framework for their analysis, Chealy et al. (2014) found that most tasks in the four science subjects require students to perform not only at lower levels of thinking but also to rote-cite formulae and apply formulaic approaches to practical exercises. This situation, according to Chealy et al., has contributed to the generally negative attitudes that students hold toward science subjects.

As described earlier, despite the lack of emphasis on science experiments in science curricula and the lack of laboratory facilities in many secondary schools, especially schools in rural areas, some facilities for experimental work and ICT-based approaches do exist. All upper secondary schools are now supplied with a standard package of science equipment that provides teachers with the opportunity to effectively deliver science content (MoEYS 2013). However, emphasis on experiment-based topics remains low because of the limited capability of teachers, the lack of laboratories and equipment, and the paucity of quality instruction manuals. In essence, almost no experimental work takes place during the teaching and learning of science subjects. A Volunteer Service Overseas (VSO) study indicates that this situation applies equally to the lower and upper secondary levels of schooling (VSO 2014).

Full understanding of the reasons for the low quality of student learning in science subjects remains difficult because Cambodia has no national assessment test at the upper secondary level and also because science teacher performance at the high school level has not been subject to careful research scrutiny. Nonetheless, some Cambodian students who are outstanding in physics and mathematics have been awarded medals and awards by the Regional Congress Search for SEAMEO Young Scientist and in some International Olympiad competitions.

Continuing Professional Development Issues

As mentioned earlier, teachers have almost no opportunities for continued professional growth. Although the most recent Education Congress Report (2013) lists nearly 50 inservice programs, these reach only a tiny proportion of teachers and are not organized around a unified curriculum or vision. There is little incentive for teachers to participate in INSET because professional development is not linked to promotion or a longer-term vision for teacher career growth. It remains ad hoc, inconsistent, and unmonitored. An institutionalized INSET system needs to be put in place and to be accompanied by a credit scheme ensuring equivalency of training and training awards along with incentives such as promotion, increased pay, and the awarding of advanced qualifications. Most importantly, all such measures need to be linked to a teacher career development pathway (MoEYS 2015).

Cambodia's inspection process also needs mentioning. The country currently has two types of inspection: school internal inspection and external inspection. The goal of teacher inspection is to observe the pedagogical practices that teachers use in class, including how they use and deploy textbooks and teaching materials and how well they are able to motivate students. This form of inspection is mostly accomplished through class observation and follow-up interviews. Teachers do receive feedback on their performance, with that feedback intended to provide them with direction on future improvement (Choun 2014). However, the application of a robust assessment method such as Bloom's Solo Taxonomy is more likely to lead to changes in teaching methodology such that teachers move away from the traditional transmission-based pedagogy of teaching and toward more scientific inquiry-based pedagogies.

Trends and Developments in Upper Secondary Science Teacher Quality

The Royal Government of Cambodia has a clear vision to raise Cambodia from a lower-middle-income country to an upper-middle-income country by 2030 and a developed country by 2050. In response to this vision, the MoEYS is playing an important role in producing high-quality human resources to meet labor-market demand. The Education Strategic Plan 2014–2018 (MoEYS 2014) has accorded significant priority to equitable access to high-quality basic education services. A particular target is that of increasing the quality of teachers at the lower and upper secondary levels of the school system. These developments will be effected through a program focused on developing teaching and learning materials; upgrading preservice and inservice training programs for teachers and school administrators; reforming the assessment system, including the national examinations; and developing infrastructure and facilities, including labs and computer rooms.

In order to achieve these goals, the MoEYS needs to effectively mobilize resources. This process requires prioritizing human resource development with respect to both science and education experts so that they can help educate and train high-quality science teachers and produce the teaching and learning materials and environments needed to equip students to meet the social and economic needs of the twenty-first century. A robust, high-quality inquiry-based curriculum within the Cambodian context, supported by materials for teaching and learning science, are the tasks that the MoEYS most urgently needs to address.

Building up a twenty-first-century science teacher education model and providing ongoing professional development opportunities to all science teachers in Cambodia are critically important. The MoEYS needs to change its current practice of awarding this work solely to the NIE by allowing capable higher education institutions to provide teacher education programs, especially in the field of science subjects. Such a change will allow graduates with science expertise diversified entry points to training as high school science teachers.

References

Ayers, David. 2000. *Anatomy of a Crisis: Education, Development, and the State in Cambodia, 1953–1998*. Honolulu: University of Hawaii.

Benveniste, Luis, Jeffrey Marshall, and M. Cardidad Araujo. 2008. *Teaching in Cambodia*. Phnom Penh: World Bank.

Chealy, Chet, Chanrith Ngin, Sitha Chhinh, Sideth S. Dy, and David Ford. 2014. "Reviews of Educational Contents, Pedagogies and Connectivity of Curriculum and Its Relevance to Economic Development in Cambodia: A Focus on Khmer, Mathematics, and Science for Grades 9 and 12" (unpublished report). Phnom Penh: Royal University of Phnom Penh.

Chhit, L. I. 2013. "Teacher Trainees' Perceptions toward Teacher Career." Unpublished Master's diss., Royal University of Phnom Penh, Cambodia.

Choun, Naron Hang. 2014. "Reforming Institutions to Improve Education Service Delivery and Strengthen Cambodia's Competitiveness." *Analysis of Development Policy and International Cooperation* 1 (1): 121–57.

Extraordinary Chambers in the Court of Cambodia (EECE). 2010. *Judgement*. Phnom Penh: Author.

Ho, Ester Sui-cho. 2013. *Student Learning Assessment* (Asia-Pacific Education System Review No. 5). Bangkok: UNESCO, http://unesdoc.unesco.org/images/0021/002178/217816E.pdf.

Hout, M. 2013. "Textbook Analysis: Level of Critical Thinking in Cambodian Biology Textbook." Unpublished Master's diss., Royal University of Phnom Penh, Cambodia.

Japan International Cooperative Agency (JICA). 2012. *Cambodian Science Teacher Education Project*. Tokyo: Author.

Madhur, Srinivasa. 2014. *Cambodia's Skill Gap: An Anatomy of Issues and Policy Options* (Working Paper Series No. 98). Phnom Penh: Cambodia Development Resource Institute (CDRI), http://www.cdri.org.kh/webdata/download/wp/wp98e.pdf.

Ministry of Education, Youth, and Sport (MoEYS). 2004. *Policy for Curriculum Development 2005–2009*. Phnom Penh: Author.

———. 2010. *Teacher Development Master Plan*. Phnom Penh: Author.

———. 2013a. *Education Indicators and Statistics*. Phnom Penh: Author.

———. 2013b. *Education Congress: Report on the Education, Youth and Sports Performance In The Academic Year 2011–2012*. Phnom Penh: Author.

———. 2014. *Education Strategic Plan 2014–2018*. Phnom Penh: Author.

———. 2015. *Teacher Policy Action Plan*. Phnom Penh: Author.

UNESCO. 2008a. *Asia and the Pacific Education for All (EFA) Mid-Decade Assessment: Mekong Sub-Region Synthesis Report*. Bangkok: Author.

————. 2008b. *Secondary Education* Regional *Information Base: Country Profile—Cambodia*. Bangkok: Author.

————. 2010. *Current Challenges in Basic Science Education*. Paris: Author.

Voluntary Service Overseas (VSO). 2014. *Experiment Support Book for Science Teachers Grades 7–9*. Phnom Penh: Author.

World Bank 2014. *Educating the Next Generation: Improving Teacher Quality in Cambodia*. Phnom Penh: Author.

Further Reading

Khieng Sothy, Srinivasa Madhur, and Chhem Rethy. 2015. *Cambodia Education 2015: Employment and Empowerment*. Phnom Penh: CDRI.

Ministry of Education, Youth, and Sport (MoEYS). 2015. *Education Congress: Report on the Education, Youth and Sports Performance in the Academic Year 2013–2014*. Phnom Penh: Author.

UNESCO. 2013. *Policy Review of TVET in Cambodia*. Paris: Author.

21

Uganda

Charles Kivunja and John Sentongo

Teaching Science in Uganda

Backdrop

Nestled among the Great Lakes of Edward, Albert, Kyoga, and the world's largest freshwater lake, Lake Victoria, and bestriding the equator, on a plateau over 1,000 meters high, traversed by the River Nile, landlocked Uganda is a natural wonder of beauty that truly deserves the label "The Pearl of Africa" that Sir Winston Churchill gave it in 1909. Churchill must have been right because the Lonely Planet characterized Uganda as the number one tourist destination in 2012.

Uganda's population data vary depending on which source one consults. Two that concur are Coutrymeters (2015) and the World Population Review (2014). The Coutrymeters' data, which appear to be the most up to date, estimated that the population (as of March 11, 2015) was 40.6 million. This estimate aligns well with that of the World Population Review, which reported a total population for 2014 of 39.2 million. The review also estimated the population to be growing at 3.6 percent per annum, a rate that would likewise give a figure of over 40 million people in 2015.

According to the World Population Review (2014), 48 percent of the Ugandan population is below 15 years of age. Coupled with a high fertility rate of 6.9, this means that Uganda has a very young population growing at the fastest rate in East Africa. Thus, more than half of the population falls within the secondary and university education age bracket commonly taken to be between 15 and 24 years of age. At independence in 1962, the country had a population of 7.2 million. But even with the wars that led to loss of lives in large numbers, the population has more than quadrupled between 1962 and 2015. The high fertility rate is attributed to low levels of education, poor access to family planning services, a low rate of use of contraception strategies, early childbearing (with an estimated 25 percent of adolescent females getting pregnant before they turn 19), cultural preference for large families as a source of security in old age, and religious beliefs. Average life expectancy is 52.7 years, with females living longer than males. Women have a life expectancy of 53.8 years; males, 51.7 years.

Uganda has a very rich mix of ethnic groups, but on the whole the different tribes get on quite well with one another, and over the last 50 years the situation has been one of harmonious cohabitation. The four largest ethnic groups are the Baganda, Banyankole, Basoga, and the Bakiga, who together make up 42 percent of the population. Uganda is home to another 13 tribes, but they each represent relatively small percentages of the total population. Most Ugandans are Christians; Muslims make up 12 percent of the population.

The country has made reasonable progress toward its goals of universal primary education and eradicating illiteracy. Access to universal primary education increased from 2.5 million in 1997 to 7.5 million in 2008, and the enrolment percentage has reached 82 percent of students of primary school age. In 2007, the government introduced universal secondary education as well as universal post-primary education, resulting in an increase in secondary school enrolment by 25 percent in 2008 (Government of the Republic of Uganda 2015). The government set itself the millennium development goal of a net enrolment ratio of 100 percent by 2015, up from 84 percent in 2005/2006. The adult literacy rate increased from 69 percent in 2005/2006 to 73.6 percent in 2009. With a Human Development Index (HDI) of 0.514 in 2009, Uganda ranks 157 out of the 182 countries that compile HDI data (United Nations Development Programme 2007).

Uganda's development has experienced a variety of approaches. From 1997 to 2008, the country was guided by what was called the Poverty Eradication Action Plan. The government's current national development plan for 2010 to 2015 is designed to guide economic policies and activities for the period 2010/2011 to 2014/2015. The aim expressed in this plan is to transform Ugandan society from a peasant to a modern and prosperous country within 30 years, and the government explicitly spells out the objective of "promoting science, technology, innovation and ICT to enhance competitiveness" (Government of the Republic of Uganda 2015, 5). In regard to the current planning period, the government stated that investment priorities will include physical infrastructure development, mainly in energy, railway, waterways, and air transport; human resources development in the areas of education, skills development, health, water, and sanitation; facilitating availability and access to critical production inputs, especially in agriculture and industry; and promoting science, technology, and innovation.

After the introduction of the Poverty Eradication Action Plan, the Ugandan economy experienced varying growth rates, with an average GDP growth rate of 7.2 percent between 1997/1998 and 2000/2001 and 6.8 percent between 2000/2001 and 2003/2004, increasing to 8 percent over the period 2004/2005 to 2007/2008. Based on economic forecasts, GDP growth rate over the national development plan period is projected to reach an average of 7.2 percent per annum. At this rate of growth, nominal per capita annual income is projected to increase from US$506 in 2008/2009 to about US$850 by 2014/2015. During the same period, the proportion of people living below the poverty line is expected to decline from the 31 percent evident in 2005/2006 to about 24.5 percent in 2014/2015, which is above the millennium development goal target of 28 percent.

Formal education in Uganda is overseen by the Ministry of Education and Sports (MoES). The MoES has a number of agencies, including the National Curriculum Development Centre, which is responsible for developing school curricula for schools, the Uganda National Examinations Board, which administers national examinations to assesses students' academic performance, and the Directorate of Education Standards and the National Council for Higher Education, charged with ensuring quality in education provision at school and higher institution levels, respectively.

Uganda's basic education is divided into two stages—the pre-primary (also called nursery or kindergarten) and primary stage. The pre-primary stage caters for children 2 to 5 years of age. These schools are exclusively run by private institutions and individuals. The primary stage encompasses 7 years and caters for children 6 to 12 years of age (P1 to P7). This stage is provided by both private- and public-sector schools, with the public schools comprising nearly 75 percent of all primary schools. This is the stage when Ugandan children are introduced to science.

The full core curriculum of Uganda's primary education is easily remembered by the acronym MESS, representing the four subjects of mathematics, English, science, and social studies, which all students must complete. When the government introduced universal primary education in 1997, it made primary education free for all children, a development that led to primary enrolments more than trebling from 2.5 million in 1996 to 8 million in 2008 (Uganda National Council for Science and Technology 2010). However, the dropout rate has remained very high, with retention estimated at only 30 percent (MoES 2010). On completion of their primary learning stage, students sit their first major national examination, which is administered by the MoES. Successful graduates are awarded the Primary Leaving Certificate of Education.

The primary stage is followed by four years of lower secondary education (S1 to S4) and two years of upper secondary (high school) education (S5 to S6). Uganda's secondary education is provided by both public and private schools as well as a few international schools, which deliver foreign curricula. The majority of secondary schools (62 percent) are in urban areas with the remainder (38 percent) located in rural areas.

Uganda's education structure is modeled on the British education system. Until 1972, the country's public examinations were set and marked in Britain, and the academic award after four years of secondary education was the Cambridge School Certificate. Although Uganda still follows the British system and structure of education, all examinations are now set and marked within the country and hence the certificates students obtain are national certificates.

At the end of S4, students sit their second major national examination known as the Uganda School Certificate of Education or simply O-Level examinations. O-Level secondary education was made free in 2007 to increase access to this level of schooling, but only students who achieve a credit or better grade in these examinations (from pass, credit, distinction, high distinction) qualify for government funding to undertake it. On successful completion of S4, students are awarded Uganda's School Certificate of Education and can then advance to the Higher School Certificate stage (S5 and S6). At the end of S6, students sit their third set of major national examinations, for the Uganda Advanced Certificate of Education. Students awarded this certificate qualify for entry to university studies. The government funds 4,000 university scholarships every year on a highly competitive basis.

The combined universal primary education/universal post-primary education strategy is estimated to have increased the gross enrollment ratio at the secondary school level from 25 percent in 2006/2007 to 28 percent in 2007/2008. For the period 2006 to 2009, the introduction of universal secondary education in 2007 is estimated to have increased secondary school enrollment (S1 to S6) by 30 percent. The transition from S4 to S5 improved from 48 percent to 51 percent for the 2008/2009 to 2009/2010 period (Uganda National Council for Science and Technology 2012). Nevertheless, teachers in many classrooms do not have the number of students in their classes they could accommodate because in general the number of students attending secondary education is still very low. For example, in 2009 the net enrollment ratio for students in the secondary school age bracket of 13 to 19 years of age was only 24 percent (MoES 2010). These data suggest that many Ugandan young adults of secondary school age who are supposed to be at school are actually not attending the formal school system. This non-compliance not only frustrates the government's efforts to achieve universal secondary education but also undermines national efforts to see many more students enrolled in science education.

In 2009, 65,000 teachers were teaching at the secondary school level in Uganda, and the secondary school student population at that time was 1.2 million enrolled in both public and private secondary schools. Secondary school teachers in Uganda fall under two categories: Grade V teachers, who have completed six years of secondary education and a three-year diploma course in education, and graduate teachers, who are the majority and have first degrees or Master's degrees; a few have doctorates.

After high school, most students join institutions of higher learning depending on their academic performance in the Uganda Advanced Certificate of Education examinations and their interests. The first choice is invariably a university, given the academic status that it represents. Uganda presently has six public and over 20 private universities, and it is this sector of the education system that is primarily responsible for training teachers, including, of course, science teachers. The normal undergraduate degree is completed over three years in a university. Other categories of post-secondary institutions include national colleges of commerce, national teachers' colleges, and technical colleges and other departmental training institutions that offer specialized training. O-Level leavers who do not proceed to high school can join primary teachers' colleges, technical institutions (including business, technical, and vocational education and training), farm schools, or other departmental training institutions. All students who join institutions of higher learning other than a university can acquire university education through upgrading. Their college qualification, a two-year diploma, is considered the equivalent of A-Level.

School Science

The government makes a direct link between science education and the country's capacity to produce scientists who will improve Uganda's economic development and Ugandans' standards of living. In its latest national development plan (2014/2015), the government posited science education as one of the key pillars to support the attainment not only of its millennium development goals but also several of its national development goals, through its formative role in the supply of professionals such as doctors, engineers, agriculturalists, and foresters. These ambitious goals need to be contextualized with the understanding that most Ugandans, particularly in the country's rural and remote areas, are currently scientifically illiterate.

Thus, Uganda considers the teaching of science to be a strategic investment for the country in terms of training the critical mass of scientists and engineers needed to spearhead its economic growth and development efforts. As emphasized in the national development plan, the quality of science education is determined by the quality and number of science teachers, the teaching infrastructure, and the relevance of the curriculum to the needs of the country. The plan thus asserts that the level of enrolment and performance of science students at school is a reflection of the quality of science education. Accordingly, in 2006, the government introduced a policy that made the core science subjects of physics, chemistry, biology, and mathematics compulsory for all lower secondary students (Uganda National Council for Science and Technology 2012).

Currently, all students at O-Level study chemistry and physics, unlike before when these two subjects were optional after S2. Mathematics, English, biology, and geography have always been compulsory at this level. To ensure good grades in national examinations and minimize running costs, many private schools avoided chemistry and physics and concentrated on arts subjects, which they claimed were easier for students to pass.

Separate science disciplines are introduced in the first year of O-Level (i.e., S1). The science disciplines provided at this level include biology, chemistry, physics, and agriculture. The first three subjects are compulsory; the fourth is optional. Despite the fact that Uganda is an agricultural country, agriculture as a subject has never been compulsory at any level of education. Consequently, the subject attracts very few students. The compulsory status of the three science subjects is part of the argument for prioritizing the teaching of science in pursuit of scientific and technological development and industrial growth. All schools except international schools follow the same O-Level syllabus, which is authored by the National Curriculum Development Centre, and take the same Uganda School Certificate of Education examinations.

The O-Level syllabus is designed to meet the needs of students who intend to pursue science disciplines at A-Level as well as those who leave school after O-Level for the world of work (National Curriculum Development Centre 2008). The syllabus document presents science disciplines as practical subjects, where science concepts, principles, and skills are supposed to be developed through experimental investigations by the discovery learning method. During the four years of O-Level, students' academic performance/progression is determined through both formative and summative assessment.

The A-Level science syllabuses are intended for students who wish to pursue science- related careers and so require deeper understanding and application of science concepts and principles. A-Level syllabus content continues on from that of the O-Level curriculum. The syllabuses suggest the order in which content should be taught, teaching and learning strategies, and assessment strategies. In all subjects, the syllabuses express the need to integrate theory with practical work. As is the case at O-Level, students' academic performance at the end of A-Level is assessed by summative means only. Formative assessment is school based and does not contribute to national examination marks. In all the science disciplines, students take the same number of examination papers at the end of each course—that is, two theory papers and one practical paper; the difference resides in the duration and composition of the papers.

Experience over the years has shown that Uganda's school system is effective in preparing students for further studies. High-school leavers qualifying with the Uganda Advanced Certificate of Education can undertake undergraduate studies wherever they choose. However, a close examination of the system reveals a number of policy and pedagogical issues. Over time, the Government of Uganda has introduced many changes, some of which have not been informed by research data. For instance, in 2007, chemistry and physics were made compulsory for all students at O-Level. However, many secondary schools, especially privately owned ones, did not have the capacity to effectively teach science subjects: they lacked laboratories, instructional materials, and qualified science teachers. Earlier, the government's introduction of the Uganda Certificate of School Education led to large classes at O-Level, with no corresponding improvement in both physical and human resources.

Student enrollments in science in upper secondary levels and their performance in science are commonly used as indicators of the quality of science education at this level. Unfortunately, both the numbers of students enrolled in upper secondary school science subjects and their performance in the O-Level School Certificate and the A-Level Higher School Certificate indicate that there is much room for improvement, particularly with regard to teacher quality. Students' performance in the four core science subjects of physics, chemistry, biology, and mathematics in the O-Level School Certificate of 2007/2008 was very poor, with the majority of candidates failing all four subjects. Students' performance was worst in the laboratory-based subjects, especially chemistry and physics. Commentators argued that in addition to the lack of science equipment in the schools, the most important factor contributing to the very high failure rates was teachers' instructional methods, which were primarily theoretical in implementation even in the subjects with a high practical component (Uganda National Council for Science and Technology 2012).

Teacher Academic and Professional Education and Training

The Uganda National Academy of Sciences based at the leading university in Uganda, Makerere University, was formed in 2000 and charged with the responsibility "to contribute towards improving the prosperity and welfare of people of Uganda by promoting, generating, sharing and utilizing scientific knowledge and information and to give independent, evidence-based advice to government and society" (Uganda National Academy of Sciences 2015, 6). The academy is seen as a key contributor to the academic and professional education and training of science educators in the country.

In Uganda, individuals wanting to qualify to teach school science at the high school level must sit the A-Level examinations and obtain the Higher School Certificate qualification with at least two principal passes. These individuals can then enroll for a Bachelor of Science with Education degree course, Bachelor of Science degree followed by a postgraduate diploma in education, or a diploma at a national teachers' college and then the Bachelor of Science/Bachelor of Education to upgrade to a graduate teacher.

Regardless of the route an aspiring science teacher takes, he or she has to study science subject content and science education pedagogical content along with psychology, curriculum, and educational foundations and management—all essential courses in teacher education. The science subject content offered at teacher training institutions helps teacher trainees understand school science better and, at the same time, achieve a level of science-based knowledge in advance of the level they will teach their students on completing their training. The science education courses are also expected to equip trainee teachers with skills to deliver the subject matter effectively. However, experience shows that teacher trainees find it difficult to obtain the requisite knowledge and skills because the Faculty of Science lecturers who teach them generally have no pedagogical training. In addition, the trainees are overloaded with science content beyond that which is relevant to secondary school science. These aspects of the training present science teachers with challenges when delivering the science content to learners—a weakness school administrators frequently point out.

While on training, science teacher trainees are required to major in one of the two science disciplines and offer other science disciplines as a minor. Under ideal situations, newly qualified science teachers are expected to teach the discipline they majored in to both O- and A-Level students and to teach their minor at O-Level only. Although this expectation accords with MoES policy, it is often not applied in schools with a scarcity of science teachers.

Although universities are the main stream through which science education is provided to Uganda's teachers, some academic and professional education and training is provided for science education through business, technical, and vocational training (BTVET) institutes. Their role in providing teachers with certificate and diploma training programs was significantly increased by government policy measures introduced in a white paper on education and the Education Sector Strategic Plan 2003–2018. The latter provided for a system of registering and licensing BTVET institutes so they could offer recognized training in the technical skills, including science, needed in the labor market, starting from 2003.

Unfortunately, BTVET institutes receive very little funding from the government. For example, in the 2009/2010 financial year, their share of the education sector budget was only 5.5 percent (Uganda National Council for Science and Technology 2012). Nonetheless, in 2004, the Ministry of Education established the Uganda Vocational Qualifications Framework Secretariat through which standards, curricula, and assessments conducted by the BTVET institutes were established. Since then, BTVETs have made some improvement in the delivery of science education. However, BTVETs are in critical need of qualified instructors, especially in science subjects. The shortage of qualified instructors has led to poor results in science education. For example, the Uganda National Examination Board in 2008 noted high failure rates in the certificate and diploma examinations conducted at these institutes, particularly in science subjects.

The government's current national development plan provides explicit plans to improve teaching through the accelerated recruitment of more qualified teachers and the enhanced preservice and inservice training of teachers. The need to train untrained licensed teachers receives particular emphasis in the plan, as does training of science teachers. The plan also sets out means of recruiting more tutors to train teachers in teacher training colleges and advises how the government intends to improve teachers' working conditions. The plan furthermore aims to use training and continuous professional development (CPD) to improve preservice and inservice secondary school teachers' teaching competency. The intention is to ultimately provide inservice training for 400 teachers annually, many of whom will be science teachers.

Continuing Professional Development

Although Uganda has an oversupply of teachers in social science subjects, there is an acute shortage of science teachers, a situation accentuated by the fact that there is minimal support for CPD. In the main, the training of science teachers is seen as a one-off process, with that perception based on the premise that, once trained, any teacher of whatever discipline is trained for life and so needs no subsequent inservice training. Thus, teachers lack the CPD required to refresh their skills and knowledge from time to time. Science teachers are victims of a system that provided them with poor preservice training, does not support them with adequate secondary science teaching infrastructure, and does not provide ongoing professional support.

That criticism aside, the MoES is providing science and mathematics teachers with some measure of professional development through a nationwide Secondary Science and Mathematics Teachers (SESEMAT) program. Its aim is to show teachers how to use constructivist principles in the teaching and learning process. The MoES organizes SESEMAT activities during school holidays, and they are residential. The program resulted in 2,500 secondary science teachers trained by 2008 and an increase in the supply of science curriculum materials. There have also been reports of improved teacher attitudes toward the teaching of science subjects. However, there is, as yet, no evidence of improved student performance in secondary science subjects (Uganda National Council for Science and Technology 2012).

Apart from SESEMAT, there is no other national CPD initiative. To fill the gap, most secondary schools conduct school-based seminars and/ or workshops for their teachers. In addition to SESEMAT, the seminars, and workshops, many teachers take personal responsibility for their own professional development by enrolling in different academic programs leading to certificates, diplomas, or higher degrees. The common qualification for science teachers is the Master of Education in science education.

Science Teacher Professional Associations

Uganda's science teacher professional associations can be traced back to the formation of the Uganda Teachers Association in Mbale in the 1940s. Established as a vehicle for presenting teachers' grievances to the then Department of Education, the association became the most recognized professional body for teachers, and by independence in October 1962 most of Uganda's secondary science teachers belonged to it. Primary teachers were reluctant to join the association and so formed their own called the Uganda Primary and Junior Secondary Union. It was more of a trade union than a professional association.

The Uganda Teachers Association grew its membership, and by 1971, 50 percent of all teachers in Uganda were members. The association represented teachers on most decision-making bodies dealing with matters affecting education in general and teachers in particular. It made arrangements for teachers to attend inservice courses, organized life insurance for them, and achieved a unified status for all teachers as employees of the Uganda government. The association also continued to provide a channel through which teachers' grievances and suggestions could be heard by the government. The Uganda Teachers Association presented itself as a truly nonmilitant professional organization, with a well-articulated code of conduct and philosophy.

Many members of the Uganda Teachers Association are also members of Uganda's National Association of Science Teachers, which was formed in 2010 to specifically cater for the interests of science teachers and promote excellence in science teaching and learning at all levels throughout the country. The National Association of Science Teachers is affiliated with the Uganda National Academy of Science and the Uganda National Council for Science and Technologies, as well as some international science organizations. One of the National Association of Science Teachers' focus areas is the education of girls in science. The association also works with the MoES to support the science teaching profession by developing and reforming science curricula, preparing examinations, aiding teacher training, and facilitating best-practice science pedagogy. The association furthermore works with Makerere University to encourage professionalism in science teaching by providing moral and practical support such as conferences, workshops, and laboratories for teachers to train in and to practice setting up experiments. Unfortunately, the National Association of Science Teachers does not have a high profile; most science teachers in Uganda are not aware of it.

Issues in Upper Secondary Science Teacher Quality

Teacher Supply

There is an acute shortage of science teachers in Uganda, especially in the rural and remote areas. The situation in rural areas is appalling, as most teachers do not want to stay in these places. The MoES posts teachers to the rural schools, but some of these teachers do not even report to their designated school, while the majority of them work in these schools for a day or two and then return to urban centers to spend most of their time in the schools there. The shortage of science teachers in rural areas is therefore acute, and especially so in the small private schools, which cannot afford to employ well-qualified full-time teachers.

One of the reasons for the shortage is that students specialize in their career paths relatively early, when they graduate from senior secondary schooling, at which stage many do not have the knowledge and skills to consider a career in science. Enrollments into tertiary-level science subjects and science teaching are therefore low. The problem of teacher shortages starts at primary school levels and continues to the upper secondary levels because many schools do not have science-teaching role models. The gross enrollment ratio of students into science subjects is only 1.5 percent. Of this dismal percentage, only a few students successfully complete their course. This means that students in these schools do not choose science subjects, a choice that feeds into the choice of becoming a science teacher. The shortage of science teachers is also greater among female than male teachers.

Unsurprisingly, many of the teachers teaching science in secondary schools do not have training in science subjects. For example, an MoES report in 2012 found that 14 percent of teachers in secondary schools did not have the pre-scribed minimum qualifications for teaching in a secondary school. Of those qualified, only a third had a Bachelor's degree in science education. According to the Education and Sports Sector Annual Performance Review 2009/2010, the shortage of qualified science teachers is so acute that most science subjects are taught by so-called "A-Level dropouts." These are people who reached S6 but failed the Uganda Advanced Certificate of Education and could not gain entry into university. Moreover, many classrooms do not have adequate sci-ence equipment and resources. This trifecta of a shortage of trained secondary science teachers, the poor quality of those teaching secondary science, and the lack of necessary materials has resulted in a conundrum involving teach-ers who do not have the expertise needed to teach secondary science and are unable to improvise materials and resources needed for an effective science lesson, frustrated secondary students enrolled in secondary science subjects, and poor academic results for these students. Furthermore, as the Uganda National Council for Science and Technology (2012) notes, secondary science teachers are not adequately motivated and this, too, affects the quality of their teaching.

Despite the commitment to science education in the national develop-ment plan, the numbers of students enrolled in science subjects in the teacher training institutions and national teachers' colleges are on the decline. One of the reasons for this decline is that teachers are among the worst-paid public servants in the country, and so the profession does not attract good students. Given the high cost of living in urban areas, where many science students are enrolled and where the majority of science teachers work, science teachers often have to teach at more than one school just to earn enough to support their families.

Professional Education and Training Issues

The National Council for Higher Education is the body charged with accredit-ing all institutions of higher learning and also the programs these institutions offer. After receiving accreditation, each institution trains its teachers accord-ing to its own prescribed program. Although there are some similarities in the teacher training programs offered by the different institutions, they are not uniform except for two components—theory, conducted at the institution, and the practicum, carried out in schools for periods ranging between 12 and 24 weeks. These differences mean that trainees from different institutions graduate with varying teaching competencies, hence some school adminis-trators, especially those for private schools, exhibit preferences for graduates from some institutions and not others. Experience shows that science gradu-ate teachers are generally very strong in terms of science content but weak when it comes to delivering that content to learners.

Because training is conducted almost entirely by the trainers of the respec-tive institutions, the institutions rely on external people to assess and assure the quality of their trainee teachers. The external examiners' assessment of the trainees' teaching competencies provides institutions with needed quality feedback. School education boards discuss the examiners' reports and then adopt any recommendations they consider worthwhile.

As part of its efforts to improve the training and professional education of science teachers and other science professionals, the Uganda government introduced a requirement that students receiving government scholarships for their tertiary studies must include science and technology in their study programs. This policy was aimed at changing professional and social attitudes toward science, which historically have overlooked science subjects and instead focused on the arts and social sciences.

According to Uganda's current national development plan (2014/2015), student enrolment in science and technology throughout the country, in both private and public universities, is less than 27 percent of all registered students. The plan therefore contains a government-set target of improving enrolments in science education and training in public and private schools, tertiary institutions, and universities so that the ratio of arts to sciences improves from 5:1 to 3:1. However, funding remains a major problem, with preservice teacher education especially remaining chronically underfunded.

Issues at the Chalkface

The secondary school curriculum is very crowded; students at O-Level in some schools are studying up to 20 subjects. Overcrowding means that there is only a limited amount of time for teaching and learning each subject. This time constraint becomes more critical with respect to teaching science subjects because these subjects, by their very nature, require that students be given the opportunity to cover not only the theoretical aspects of the subjects but also to complete the practical components of each one.

This problem is compounded by insufficient science equipment. Students have to share equipment and materials, take turns at doing activities (assuming there is time enough for this rotation method), or work in large groups that do not give each student sufficient opportunity for the hands-on practical experience that these subjects require. These outcomes limit students' understanding and appreciation of the links between the theoretical concepts they are taught and the practical applications of those concepts. Thus, time and resources are major handicaps to effective science teaching at the upper secondary school levels in Uganda.

The MoES has put a lot of money into improving teacher education and providing schools with the equipment and materials they need to improve the quality of teaching and learning science. However, although significant improvements have been made, the ongoing shortage of well-trained science teachers in science classrooms cannot be discounted, and most science teachers still do not have science laboratories that are sufficiently equipped to aid effective teaching of science. These problems are more pronounced in private than in public schools and are worse again in the rural and remote areas of the country. The textbooks used for teaching science are from overseas, and this makes it difficult for teachers to apply some of the materials or content to local conditions, thus making the teaching of science even more difficult. Furthermore, the theoretical methods that teachers use to teach students the pure sciences do not utilize the digital technologies that today's students find appealing and fascinating (Kivunja 2014).

At the high school level, the Uganda Advanced Certificate of Education, which gains students entry into university courses, tends to dominate teaching and learning. The curriculum is examination driven, which in turn influences the way secondary school teachers teach. Teachers are inclined to emphasize the achievement of high grades and so promote to students those subjects they are most likely to get high grades in and thus be able to attend university. Science subjects are generally presented as "difficult" subjects.

In an effort to increase the number of Ugandans with secondary science education and to improve science literacy, the government, as pointed out earlier, introduced a policy in 2006 that required all lower secondary-level students in S1 to S4 to include the core science subjects of physics, chemistry, biology, and mathematics in their study programs. However, this policy did not have the desired outcome due to high dropout rates. For example, data for the Uganda School Certificate of Education examination for 2010 show that of the cohort that had been forced to take these science subjects (starting from the enrolments in S1 in 2010) only 39 percent were left (Uganda National Council for Science and Technology 2012). The reasons given for the high dropout rate included students' and parents' perceptions that these science subjects are the ones most difficult to pass and that enrolling in them would therefore put students' chances of university entrance at risk.

School administrators in Uganda place considerable emphasis on the need to complete subject syllabuses. Most Ugandan schools assess teachers' performance in terms of syllabus coverage. Because practical work requires considerable preparation and execution time, teachers tend to concentrate on teaching science concepts theoretically so as to complete the syllabus as early as possible. In addition, the majority of school administrators do not have a science background and therefore do not adequately provide the necessary materials requisitioned by science teachers. Some administrators do not surrender funds to heads of science departments but instead buy the materials themselves. In some instances, they buy substandard materials, which do not give the expected results. Many schools in rural areas do not have enough instructional materials to conduct practical work. In some of the universal secondary education schools that do have the necessary instructional materials, teachers do not use them because they (the teachers) lack the competence to conduct experiments.

It is dangerous to generalize, but the lack of science teachers and resources needed for science education is well demonstrated by a situation witnessed at one of the secondary schools in a rural village called Kabwoya in western Uganda. This school, a new school, opened in 2008, and two American student volunteers chose to teach high school science at it. The teaching of science at the school was set up as a collaborative project between Northwestern University in Illinois, the United States, and the secondary school. On arrival, the two American student teachers found that Kabwoya was the first senior secondary school to be built in the region.

At the time the two student teachers joined the school, no student had successfully passed the examinations and so none had been allowed admission to university. The school had insufficient teaching staff, textbooks, and other classroom equipment needed to teach science effectively. The American students soon found that they had to improvise resources, organize how to source textbooks, and find out how they could procure the equipment needed to conduct science classes. They had to write the programs to be followed in the study of science, raise the funds necessary to purchase classroom materials, and pay for their accommodation, living expenses, and their round-trip air tickets from the United States.

The American student teachers found that enrolments in secondary science subjects were low, mainly because parents could not afford to pay the school fees. The situation was more acute for girls because parents preferred to support their sons' schooling over that of their daughters. They also found that the predominant mode of instruction was the traditional lecture method. They tried to introduce a more participatory hands-on approach, but their efforts were hampered by the lack of equipment and necessary resources. They managed to source some equipment from Engineering World Health but were still far short of what they needed for effective teaching. They tried to teach the Ugandan students how to build low-cost tools and use them to aid their communities, but this was no easy job because it meant students had to learn and do community service work simultaneously. In addition to teaching school science, the American students also tried to teach about real-life health issues such as HIV/AIDS, malaria, and other infectious diseases, to equip the Ugandan students with the local knowledge needed to prevent the spread of these diseases (Massa and Wofsey 2015).

Despite the difficulties associated with teaching science, Uganda's national curriculum documents (MoES 2008) stipulate that teachers teaching science, like any other teachers, should emphasize student-centered learning. The documents also stress that teaching should emphasize the development of the competencies, life skills, and values that are essential for the development of Uganda as a democratic society. In particular with regard to teaching school science, the national curriculum documents advocate the following practices and skills: application of logical thinking and analysis, conducting investigations logically, ability to apply the knowledge and skills that have been learned, ability to interpret results of experiments, critical and supportive listening, problem solving, and cooperation with others.

The documents furthermore emphasize that effective teaching relies on good class management and a classroom environment that has the necessary physical conditions, materials, activities, and routines that facilitate student engagement with learning. Teachers, according to the documents, should make sure they plan their science lessons well and are in control of student activities during science lessons, particularly at the point at which students complete their experiments. Teachers also need to use time efficiently so that all students can complete the experiments and all other class work within the time allotted, and they must be prepared to delegate and to promote leadership among students, particularly when science experiments are being conducted in study groups. Science teachers must furthermore make sure that children remain on task all the time and that they behave in a manner that promotes positive on-task behavior. Heed must also be paid to individual student differences. Teachers additionally need to develop supportive relationships among students and encourage students to take responsibility not only for their learning but also for the learning environment, such as keeping the science laboratory clean and safe.

Continuing Professional Development Issues

As already observed in this chapter, little or no funding is available for inservice teacher training programs in Uganda. What little there is comes from non-government agencies and foreign donors. This lack of CPD support has also contributed to the shortage of science teachers. However, the government is committed to putting money into curriculum reforms, and science is one of the target areas for improvement. The MoES plans to ensure that its National Curriculum Development Centre tutors regularly observe teaching and learning so as to continuously help teachers improve their instructional methods and management. The ministry wants these tutors also to carry out orientation and inservice training of teachers in both science and ICT with the aim of enhancing teaching and, from there, student learning.

Trends and Developments in Upper Secondary Science Teacher Quality

Due to the nationwide outcry over poor performance in science in Ugandan high schools, stakeholders have been claiming that the country's teacher training institutions are the main culprits with regard to this problem. Training institutions are therefore being urged to review their programs. In response, training institutions are conducting thorough investigations intended to lead to the development of teacher training programs informed by research data. This approach is likely to improve the quality of science teachers.

The Uganda National Academy of Sciences has made some plausible recommendations, which have the potential to improve both the supply and quality of science teachers in upper secondary schools. The recommendations include students not being permitted to specialize in their subjects or career paths before they complete the O-Level School Certificate of Education in S4. If students were to take two subjects in humanities at A-Level and the other two in sciences, this complement of subjects would better prepare them for choosing science subjects once they enter high school and begin their work toward procuring their Uganda Advanced Certificate of Education.

However, the issue is complicated by the fact that by far the majority of the 20 percent of students who enroll in science subjects at senior high school do not perform well. For example, during the period 2005 to 2008, less than 10 percent of these students passed the Uganda Advanced Certificate of Education. These trends are worse in schools located in the rural areas. In 2009, for example, whereas 50 percent of students in urban schools passed their science subjects, only 20 percent of those in the rural schools passed (Womakuyu 2009). This trend was explained by the lower number of qualified science teachers in rural schools and by the greater shortages of science equipment and resources in them. Unfortunately, because these problems are endemic to schools in rural areas, it looks as though these deteriorative trends are set to continue for the foreseeable future.

The Uganda National Academy of Sciences has rightly noted that the low rate of uptake of science is a factor retarding the supply of teachers in Uganda. It recommends increasing the rate of uptake to 40 percent of student cohorts, as this would improve not only the supply of teachers but also of scientists to industry and thereby speed up national economic growth. The academy also recommends the introduction of a robust inservice program with special emphasis on science teachers' professional development. Another concern for the academy is that the language used to teach science is too complicated and therefore needs to be simplified so that science terms have relevance in people's daily transactions and living. The academy considers that if the curriculum emphasized acquisition of skills rather than mastery of facts for rote memory recall, students would be far more likely to develop skills and competencies in science subjects. According to the academy, it would also help if the government set up a full-fledged ministry of science and technology rather than have this portfolio remain within the Ministry of Education and Sports. Finally, the academy advises, additional resources need to be channeled into science education at all levels of the education system.

References

Countrymeters. 2015. "Uganda Population 2015," http://countrymeters.ifo/en/Uganda.

Government of the Republic of Uganda. 2015. *National Development Plan (2010/11–2014/15)*. Kampala: Author.

Kivunja, Charles. 2014. "Theoretical Perspectives of How Digital Natives Learn," *International Journal of Higher Education* 3 (1): 94–109.

Massa, Sean Campbell, and Katherine Wofsey. 2015. "Teaching Science and Health in the Kabwoya Village School, Uganda, East Africa." Unpublished paper, Northwestern University, Evanston, IL, http://www.davisprojectsforpeace.org/media /view/524.

Ministry of Education and Sports (MoES). 2010. *Ministry of Education and Sports, Education Statistical Abstract 2009*. Kampala: Statistics Section, MoES, http://www

.ubos.org/onlinefiles/uploads/ubos/pdf%20documents/PNSD/2009EducStatAbst
.pdf.

National Curriculum Development Centre. 2008. *National Curriculum Teachers'
Guide.* Kyambogo, Kampala: National Curriculum Development Centre, MoES.

Uganda National Academy of Sciences. 2015. *Policy Recommendations for Improving the Teaching and Learning of Science in Uganda: Uganda National Academy
of Sciences/ Science Advisors to the Nation.* Kampala: Uganda National Academy
of Sciences, Makerere University Campus, http://ugandanationalacademy.org
/policy%20beriefs/UNAS%20Statement%20on%20Teaching%20email.pdf.

Uganda National Council for Science and Technology. 2010. *Uganda's Statistical Report,
2008/2009.* Kampala: Author. http://www.uncst.go.ug/dmdocuments/Uganda%20
STI%20Status%20Report%20for%20Fiscal%20Year%202008-2009%20%20
final%20version%20for%20publication.pdf.

———. 2012. *The Quality of Science Education in Uganda.* Kampala: Science and
Technology Policy Coordination Division, Uganda National Council for Sci-
ence and Technology, http://www.uncst.go.ug/dmdocuments/Quality%20of%20
Science%20Education%20final%20report%20Dec%202012.pdf.

United Nations Development Programme. 2007. *Uganda Human Development Report.*
Geneva: Author.

Womakuyu, Frederick. 2009. "Should Struggling Students Be Forced to Study Sciences?
A Case Study of Uganda." Doctoral diss., Maryland University, College Park, MD.

World Population Review. 2014. "Uganda Population 2014," http://worldpopulation
review.com/ countries/uganda-population/.

Further Reading

National Board for Professional Teaching Standards. 2002. "What Teachers Should
Know and Be Able to Do." Arlington, VA: Author, http://www.nbpts.org/sites/default
/files/what_teachers_should_know.pdf.

Uganda Bureau of Statistics. 2014. *Uganda Demographic and Health Survey.* Kampala:
Government of the Republic of Uganda.

UNESCO. 2005. *EFA Global Monitoring Report, 2005.* Geneva: Author.

World Bank. 2014. *World Population Review: Uganda Population.* Washington, DC:
Author, http://data.worldbank.org/country/Uganda.

22

Cluster Summary Three

We end with four countries that are seriously disadvantaged with regard to their ability to provide quality education for their citizens. They are among the poorest countries in the world by national income level and so need to make judicious use of their sparse fiscal resources. This reality manifests itself in a situation where upper secondary science provision is incomplete with respect to the requisites that science education needs, including a teaching force that is adequate in terms of quantity and quality.

Teacher Supply

Whatever the situation for teaching in general in these countries, the supply of qualified, trained upper secondary science teaching professionals tends to be well below need. While shortages may not be so visible in the main urban centers, the deficit is likely to be acute in rural areas. Quality science graduates are sought after by other government departments and by the private sector, and teaching is not a competitive option for many an ambitious graduate. The gap between supply and demand tends to be filled by poorly qualified and untrained personnel—often personifications of Shaw's "those who can't."

Teacher Academic and Professional Education and Training

As already noted, many teachers in this cluster are poorly prepared for the profession. The challenge is often one of bringing practicing teachers up to par. In systems such as that of Bangladesh, where teachers receive their "preservice" training after commencing work, the distinction between preservice and inservice training is a moot one. The aim nevertheless remains that of placing an academically qualified and professionally trained teacher in front of every class. Continuing professional development therefore often also focuses on content mastery as well as pedagogical training, even though the fiscal and resource bases for these endeavors are limited.

Issues at the Chalkface

It is in this cluster that we encounter debilitating deficiencies in the resource base needed for effective upper secondary science teaching, such as school laboratories with adequate equipment. Even where this provision is present, a question mark hangs over the effectiveness of its use when the teachers deploying it are barely subject competent and have little or no training in science teaching methods. The imposition of revised curricula incorporating foreign ideological imports of the constructivist genre strikes the current writer as being somewhat ludicrous in this context.

Science Teacher Professional Associations

These tend to be either nonexistent or poorly patronized. However, the Ugandan National Association of Science Teachers, established in 2010, is a step in the right direction.

23

Banking on Quality Upper Secondary Science Teachers: An Investment Guide

Barend Vlaardingerbroek and Neil Taylor

Recognition of the importance of upper secondary science education as a human capital development activity appears to be widespread across the countries featured in this book. Ironically, this awareness seems to be the least pronounced in some of the more affluent nations, where access to upper secondary schooling is near universal (there being little or no prospect of employment for unskilled youths) and where post-primary schooling tends to occur under the same institutional roof. In such systems, it is difficult to speak of "upper secondary teachers" as a discrete professional set because all teachers are expected to service the full range of post-primary grades. The importance of the upper secondary years is also lessened by the weakening of external examination systems that control the flow of students into the tertiary education sector. Systems with distinct, restricted-access upper secondary school tiers culminating in high-stakes external examinations that determine students' subsequent fates tend to be those systems where the profile of the upper secondary science teacher as a key professional is, or should be, at its peak.

In our introduction to this book, we posited four broad criteria for evaluating upper secondary science teacher quality: subject qualifications, preservice training, a conducive work environment, and meaningful continuing professional development (CPD). The order in which these arise is not arbitrary. Ideally, teachers of upper secondary science should be well qualified academically firstly in science and then in pedagogy and teaching methods. From there, they should be able to move into well-resourced and supportive work environments, and then embark on career-long CPD. This order is generally followed in affluent and middle-income countries, at least with respect to younger generations of teachers. In the less well-off systems, we encounter serious deficiencies with regard to one or more of these areas. The rectification of these shortcomings may involve deviating from the "perfect path" described by the foregoing sequence to focus instead on onsite pedagogical and methods training.

Academic Subject Qualifications

The past decades have seen a definite move toward raising the academic credentials of upper secondary teachers to subject-graduate level. Even in systems that continue to be populated by many teachers who are poorly qualified by today's standards, there is an expectation that fresh recruits into the profession will possess degree-level academic qualifications in science. Poorly qualified teachers are moreover being encouraged to upgrade their academic credentials, and even some less affluent countries are managing to set aside funding support for this endeavor.

An upper secondary teacher of biology, chemistry, or physics simply *has to know* his or her "stuff"; no amount of educational psychology can make up for a lack of knowledge on a teacher's part about Linnaean taxonomy, the periodic table, or Newton's laws. The question arises as to how much is enough? In our view, an upper secondary science teacher should be in possession of a degree major in the subject taught. In this regard, we look somewhat askance at BEd-type "hybrid" degrees, which we consider do not measure up, despite remaining popular even in some affluent systems, particularly those where the distinction between lower secondary and upper secondary schooling is weak.

The reality is that very few teachers anywhere, even at the upper secondary level, are afforded the luxury of teaching only the one subject that constitutes their own academic major. As we have seen in this book, teachers are often expected to be able to present at least two teaching subjects. This requirement reinforces the view that a degree in science should be a *minimum* academic qualification for teachers at the upper secondary level, as science degree programs invariably include more science than just the major. Natural combinations that arise are biology and chemistry, and physics and mathematics.

Out-of-field teaching is another reality that teachers across the national income spectrum have to contend with, particularly in the light of a shortage of graduates in physical science, especially physics, that affects many school systems in richer as well as in poorer countries. The only way to effectively address this problem area is to make teaching a more attractive career to graduates in these sought-after areas, including graduates in related fields such as engineering who may already have entered the employment market and be enticed to switch to teaching.

Preservice Training

Once the preserve of teachers' colleges, most secondary school teacher training now occurs in universities or other higher education institutions. The location of the training is perhaps less important than the variation that is encountered across the individual institutions and programs. We note with approval the moves afoot in numerous countries to impose centrally dictated standards on teacher education and training programs with a view to achieving a greater uniformity of output. This endeavor is, however, a long-term project that can run up against entrenched institutional interests as well as issues arising out of central government versus devolved government (e.g., federal versus state) demarcation issues.

Science teachers need to be trained to teach science; general pedagogical training is not enough. This science-based training includes intensive "hands-on" training in the use of the school laboratory and its equipment base—the tools of *scientific* inquiry—as a fundamental science teaching environment. Where these facilities are inadequate in many schools, teachers need to know how to use and adapt those teaching resources for science that are readily available. "Trained" teachers averse to the school lab, an issue found not only in lower-income countries, are indicative of a defective training regime.

Upper secondary science teachers need to be expert not only in teaching science but also in assessment. All prospective teachers at this level need to be fully acquainted with the intricate workings of the external examinations that will decide the fate of their students (where such exams occur). We note with both approval and apprehension that the use of school-based practical assessments that count toward students' terminating marks is on the rise. This practice puts a lot of pressure on teachers who are unaccustomed to science teaching modes in which practical work plays a large part. It also encourages malpractice such as mark inflation and may involve outright corruption. Clearly, there is a role for preservice training here in assessment methods and ethics.

An issue that is gaining in prominence is the transition from preservice training to the classroom. In both higher- and lower-income countries, the need for an induction phase as opposed to the "throwing them in at the deep end" approach is becoming increasingly recognized. A model that is slowly emerging is that of a "seamless" approach to professional development in which an induction phase straddles preservice preparation and career-long continuing professional development as part of a professional development continuum.

Work Environment

Teaching in general and upper secondary teaching in particular need to be attractive professions. They have to compete with other employment destinations and career pathways for science graduates or for students entering career-oriented majors. A cultural element is also at play here, notably the traditionally high social status of educators in Asian society. In some countries, the job security and ancillary benefits (e.g., pensions) that teaching provides may be major incentives even when the remuneration is mediocre. The combination of poor pay and limited prospects for advancement does not, however, auger well for recruiting and retaining high-quality graduates into upper secondary science teaching in the countries that most need such teachers.

To do their job properly, upper secondary science teachers need physical facilities and a resource base. As we have seen from the country chapters in this book, these physical resources are generally problematic for low-income countries. The Cambodian approach of providing adequately equipped resource centers is worthy of note in this context. Technical assistance in the form of lab assistants encourages busy teachers to make more use of lab facilities in their routine practice.

The actual classroom environment is largely of a teacher's own making. The problem of didactic teaching styles is a common theme in this book. A point that we find ourselves making repeatedly both in this volume and its antecedent is that the principal function of upper secondary schooling (at least in the eyes of students and parents, if not necessarily education academics) is to prepare students for higher studies, be it at university or as part of a career development pathway. Upper secondary students, whether in Australia or Bangladesh, aiming for admission into a competitive-entry tertiary education program look to their teachers to get them into those desired destinations; they want their teachers to equip them to do well in whatever selection processes may be in operation—usually external examinations.

In short, conscientious professional educators will do what most benefits their students. If that involves exam coaching, they will exam coach. If the examination curriculum includes cookbook verification-style experiments, then that is what they will provide. In both cases, such teachers will be regarded as "good teachers" by students, parents, and school employers, whatever the academic literature in science education says to the contrary. The issue here is largely one of curricular content and emphasis, and of the assessment regime: if practical work plays little or no role in students' terminating assessment, teachers of examination classes are hardly likely to accord it high priority.

On-the-job performance appraisal and follow-up action are sensitive issues. All systems exhibit some kind of inspectoral regime, but these tend to be ritualized and to have few real consequences in terms of improving teacher performance. However, the East Asian countries in particular have well-developed, constructive evaluation mechanisms built into their continuing professional development structures.

Several chapters in this book highlight the role of professional associations with an active interest in science education. Such associations give teachers a professional identity and public voice—a voice that may be an influential one in terms of science education policy and teacher- quality improvement.

Continuing Professional Development

CPD in a significant number of the countries presented in this book—both richer and poorer— appears to be in a state of either limbo or crisis. It is unfortunately the case that CPD in many countries has long been somewhat of a shambles, usually without set standards, poorly if at all officially monitored, and often furnished by various providers all "doing their own thing." CPD frequently finds itself having the not altogether undeserved reputation of being "time off" for teachers whose classroom practice unsurprisingly does not appear to be enhanced by the experience. Little wonder, then, that CPD is among the first items to face the axe when governments prune educational funding. Yet CPD for the less affluent countries with large populations of unqualified/untrained and/or poorly qualified/trained teaching personnel is the most practicable and cost-effective way to build up a qualified and trained teaching force. This includes inservice education in science content—a "back to school" approach to bring practitioners themselves up to scratch. Content-oriented CPD is even of value to subject-competent practitioners, particularly if it is in their second teaching subject or in areas such as combined science courses where they may be operating out of field.

The danger of reliance on CPD is that participants amass "paper qualifications" (certificates of attendance, for instance) that they can pad their portfolios with but which do not give them the wherewithal to translate their newly acquired knowledge into improved classroom practice. To be effective as a means of raising the quality of teaching where it counts—in the classroom—CPD must be relevant to the daily work of teachers, and there needs to be follow-up to ensure that what has been learned is being applied. The Japanese "lesson study" approach is a particularly pertinent innovation in this regard, and it also provides a tool for constructive on-the-job teacher evaluation. CPD additionally needs to keep teachers abreast of changes in curriculum and in assessment, especially where school-based assessments have an input into final student outcomes. Technological innovations such as ICT also need to be addressed thoroughly through inservice training provision.

Authors of some of the chapters in this book mention the pursuit of higher academic qualifications under the CPD section heading. While this is not quite what we had in mind for this section, it is worthy of inclusion (assuming there is official support for it) as a means of encouraging ambitious teachers to improve their future career prospects.

Upper Secondary Science Teacher Quality: A Checklist

Academic qualifications

- At least a Bachelor's degree major in a science
- Passes in a second science as part of a degree
- Higher minimum academic requirements for upper secondary than for lower secondary

Preservice training

- At least a one-year preservice professional education program that follows set standards

- Science teaching methods courses as well as general pedagogical courses
- Practical training in laboratory-based science teaching
- Training in science-related assessment techniques and assessment ethics

Work environment

Competitive remuneration and scope for career advancement

- Structured induction phase upon commencing work
- Adequately resourced physical facilities and technical assistance
- Challenging curriculum requiring more than just rote teaching and learning
- Constructive on-the-job evaluation processes
- Membership of professional associations active in science education

Continuing professional development

- Structured and systematic career-long CPD as a professional obligation
- Well resourced and supported
- Bridges gaps between preservice education/training and practice
- Content knowledge upgrading
- Teaching methods upgrading
- Stays abreast of curricular developments
- Systematic translation of CPD-acquired knowledge and skills into practice

Overall, we consider (on the basis of the above criteria) that the East Asian threesome presented in this book—Singapore, Japan, and South Korea—lead the field in upholding and maintaining upper secondary science teacher quality. We think that other countries, including the less prosperous, would do well to study these models carefully and then try to adapt their attributes to their own particular circumstances. Of course, these are all affluent OECD members. Some of the countries with more modest fiscal resources have a lot to offer us too. Turkey takes its upper secondary science education very seriously, and Sri Lanka shows what can be achieved on a rather meagre budget. Quality upper secondary science teaching does not have to cost a fortune.

A look at the list of countries we have presented in this book makes evident that national wealth only becomes a serious limiting factor for countries toward the bottom of the middle income range. Even then, rational planning and resource appropriation can largely make up for what may appear dire shortages to a teacher in a wealthy country. At this stage, we will take the liberty of reminding our readers that upper secondary schooling in most countries of the world is post-compulsory and selective entry. The quantity/quality trade-off that poorer countries have to engage in as a result of increased access to *lower* secondary schooling need not necessarily impede the provision of properly resourced science education at *upper* secondary level. The poorest of countries can afford to provide high-quality upper secondary science educa-tion, but only to a few.

We would counsel international donors and aid agencies planning devel-opment projects for educational improvement in middle- and low-income countries to adopt the approach we have advocated in this book. This begins with the acknowledgement that upper secondary science is a critical human capital development activity but that it is quality rather than quantity that enables it to realize its potential—and that teachers are one of the key vari-ables in the upper secondary science quality equation. People who become upper secondary science teachers need to be among the cream of the aca-demic crop; hence the profession needs to be an attractive one to quality sci-ence graduates.

Such graduates will be well qualified academically in science, but they will also need to be professionally trained in science-specific hands-on teaching methods. To impart their knowledge and skills effectively, they need to work in teaching environments with the physical amenities (complete with techni-cal assistance) that are conducive to the accomplishment of their professional goals. The formation of professional science education associations also needs to be encouraged as an additional source of support and inspiration for such teachers. These teachers furthermore need career-long CPD that focuses on strengthening knowledge and skills and importantly making sure that they know how to translate these into enhanced classroom practice. The fostering of a truly professional upper secondary teaching force and ethos is a major step in a nation's socioeconomic development, and it is this that should be a major focus of development aid efforts.

Contributors

Editors

Barend Vlaardingerbroek BSc (Auckland), BA, BEdSt (Queensland), MAppSc (Curtin), PhD (Otago) is an associate professor in the Department of Education at the American University of Beirut. He was a secondary teacher in New Zealand, Papua New Guinea, and Australia before moving into teacher education at the University of Papua New Guinea (Goroka Campus) and thereafter the University of Botswana. His principal interests revolve around the human capital formation role of school education and science education, particularly in the context of developing countries. Together with Neil Taylor, he has edited the volumes *Secondary School External Examination Systems* and *Getting into Varsity* (Cambria), and *Issues in Upper Secondary Science Education* (Palgrave Macmillan).

Neil Taylor BSc Hons (Belfast), MSc (London), MA (Leicester), PhD (*Queensland University of Technology*) is a professor of science and technology education in the School of Education of the University of New England, Australia. He has been a secondary science teacher in Jamaica and the United Kingdom and has taught science education at the University of the South Pacific in Fiji and the University of Leicester in the United Kingdom. He has edited three books for an "in context series" in science education, health education, and environmental education (Sense Publishers). His main area of research interest is education in developing and emerging nations.

Foreword

John Oversby BSc, PhD (Leeds) is an Honorary Fellow at the University of Reading and Visiting Researcher at Brunel University. He has taught sciences and mathematics at secondary-level schools in Ghana and the UK. His research interests include the role of diagrams in learning sciences and mathematics (visualization), climate change education, teachers' disciplinary content knowledge, and collaborative action research. In addition to his publications in academic journals, he has edited a book on research and produced chapters for other books. He has officer posts in the UK Royal Society of Chemistry and the British Education Research Association.

Chapter Authors (in order of appearance)

Terry Lyons BEd, PhD (New England) is an associate professor of science education in the Faculty of Education at Queensland University of Technology. He taught high school science for 15 years in Australia, Botswana, and Kiribati before undertaking his PhD at the University of New England. His research interests include students' attitudes to science, participation trends in school science, teaching in rural and remote schools, and science education in developing countries. He is co-author of a number of significant Australian reports, including the *SiMERR National Survey of Rural Education* (2006), *Choosing Science* (2010), and *Starting Out in STEM* (2012).

Lee Chin Chew BSc Hons, MEd (NUS), PhD (NTU) is a senior lecturer in the Psychological Studies Academic Group of the National Institute of Education, Nanyang Technological University. She teaches Master's-level courses in educational assessment, including psychological issues related to teaching–learning assessment. She also serves as sub-dean at the Office of Graduate Studies and Professional Learning, with responsibility for coordinating Master's programs by coursework. Her research interests are in item and test development, item response theory and its applications, computerized testing, and program evaluation.

Kim Chwee Daniel Tan BSc Hons (NUS), MSc, PhD (Curtin) is an associate professor in the Natural Science and Science Education Academic Group of the National Institute of Education, Nanyang Technological University. He is currently on the editorial boards of the *International Journal of Science Education* and *Chemistry Education Research and Practice*. His research interests are in chemistry curriculum, translational research, ICT in science education, students' understanding and alternative conceptions of science, multimodality, and practical work.

Todd M. Milford BSc, BEd, MEd, PhD (Victoria) is an assistant professor in the Faculty of Education at the University of Victoria. Before this, he was a lecturer in the Art, Law and Education Group at Griffith University in Brisbane, Australia. He has teaching experience in the science and special education classroom as well as in the online environment. Todd has been teaching at the post-secondary level since 2005, primarily in the areas of science education, mathematics education, and classroom assessment. His research has been and continues to be varied; however, the constant theme is using data and data analysis to help teachers and students in the classroom.

Christine D. Tippett BASc (British Columbia), BEd, MA, PhD (Victoria) is an assistant professor of science education in the Faculty of Education at the University of Ottawa. Chris was an engineer before she obtained her teaching degree, which influences her ways of thinking about science and STEM education. Her research interests include visual representations, science education for all students, and professional development for science educators (preservice, inservice, and informal). Current projects focus on preservice science teachers' images of engineers, early childhood STEM education, and assessment of representational competence.

Lesley de Putter-Smits MSc (Utrecht), MSec, PdEng, PhD (Eindhoven) is an assistant professor at the Eindhoven School of Education, Eindhoven University of Technology in the Netherlands. She taught as a secondary science teacher in the Netherlands before combining her work with a PhD in science education. Her interests lie in the field of STEM education, specifically active learning strategies and the creation of student knowledge (constructivism).

Jan van Driel MSc, PhD (Utrecht) is a full professor of science education at ICLON Leiden University Graduate School of Teaching, of which he has been director since 2010. Jan was a chemistry teacher and gained his PhD in chemical education in 1990. Since the 1990s, his research has focused on teacher knowledge and teacher learning. Among other work, he has conducted studies on the development of preservice science teachers' pedagogical content knowledge (PCK) and on science teachers' knowledge and beliefs in the context of curriculum reform. He was a member of the board of the National Association for Research in Science Teaching (NARST) from 2009 to 2012. Jan is also on the editorial boards of several journals and since 2009 has been an associate editor of the *International Journal of Science Education.*

Toshihide Hirano BEd, MEd, PhD (Hiroshima) is a professor in the Department of Science Education at Aichi University of Education. He began his career in primary and secondary school teacher education and training at the Faculty of Education of Shimane University. He is interested in the influences of the structure of curricula and group communications on learners' concept construction in science education and in science teacher education.

Peter Rawlins BSc, MEdSt, PhD (Massey) is a senior lecturer with the Institute of Education at Massey University, New Zealand. A former secondary school teacher and assistant principal, he researches and teaches mainly in the areas of assessment and mathematics education. He was previously Director of the Graduate Diploma of Teaching (Secondary) and is currently Director of Postgraduate Studies for the Institute of Education.

Carrol Walkley BSc, MSc (Massey) is a lecturer in science education at the Institute of Education at Massey University, where he teaches in the initial teacher education programs for primary and secondary teachers. Prior to this, he taught chemistry and biology at the secondary school level. During this time, he was also involved in examining and assessing national qualifications and with inservice teacher training.

Nam-Hwa Kang BS, MS (Seoul), PhD (Georgia) is an associate professor at the Korea National University of Education. Before working in Korea, she was an associate professor at Oregon State University in the United States. She has a physics teaching license and years of experience teaching at the middle school level in South Korea. She has been teaching preservice and inservice science/physics teachers since 2002 both in the United States and in South Korea. Her research centers on science teacher developments in inquiry teaching with an additional focus on teachers' epistemological beliefs.

Abdullah K. Ambusaidi BSc (Sultan Qaboos), MSc (Warwick), PhD (Glasgow) is a professor in the Department of Curriculum and Instruction at the College of Education, Sultan Qaboos University in Oman. He has held the posts of Assistant Dean for Undergraduate Studies and Head of Curriculum and Instruction Department and is currently Dean of Postgraduate Studies. He has co-authored science teaching methods books (in Arabic) and eight book chapters on science, environmental, and health education in Oman (in English). His interests revolve around students' alternative conceptions, inquiry-based learning, the classroom environment, and active learning of science.

Muammer Çalik BSc, MA, PhD (Karadeniz) is Professor of Chemistry Education at Karadeniz Technical University. His interests focus on student learning and instructional strategies of upper and lower secondary schooling and science education. As well as being the author of more than 60 papers, he is the author of seven chapters in edited volumes on science education in context, environmental education in context, health education in context, external examination systems, university entrance systems, and upper secondary science education.

Noraini Binti Idris BScEd Hons, MEd (Malaya), PhD (Ohio) is Deputy Vice-Chancellor of the Sultan Idris Education University, Malaysia, prior to which she was Dean of the Faculty of Education at the University of Malaya. While she was in the United States, she organized several workshops to help improve minority performance and was awarded a Distinguished Diversity Enhancement Award. She was also awarded the Graduate Research Alumni Student Award. She won a gold medal at the International Innovation and Invention Competition 2005 in Geneva.

Penelope Serow BEd, PhD (New England) is an associate professor at the University of New England, where she is a project leader of the Nauru Teacher Education Project. Penelope teaches mathematics education to preservice and inservice teachers within primary, secondary, and postgraduate programs. Her research is directed at ICT as a teaching tool in the mathematics classroom, mathematics curriculum development in developing countries, and building local teacher capacity in Pacific Island countries.

Salanieta Bakalevu BEd (USP), BEdHons, PhD (Waikato) is a senior lecturer in mathematics education at the University of the South Pacific where she teaches graduate and undergraduate courses and supervises graduate students' projects. She has more than 30 years of experience as a teacher and teacher educator. Her teaching areas are mathematics education, assessment and evaluation, and gender and education. Her research interests include the influence of culture and language on mathematics and mathematics education and development of inclusive curricula.

Mesake Dakuidreketi BSc (USP), MSc (Waikato), PhD (Canterbury) is a lecturer in science education at the University of the South Pacific. He had more than 18 years' experience as an educator at the secondary school level in Fiji, and he rose to the level of Head of Department (Science) before joining USP. He has intensively researched issues in primary and secondary school science in Fiji. His interests cover a variety of topics, such as theories and views about what science is and the nature of scientific inquiry, macro- and micro-level contexts of science teaching and learning, development of children's thinking, classroom teaching practices and interactions, ecological and sociocultural theory of development and learning, and literacy and science.

Temalesi Maiwaikatakata BSc (USP), MSc (Curtin) is an assistant lecturer in science education at the University of the South Pacific. She started her career as a secondary school teacher and taught basic science, biology, and chemistry at different levels in secondary schools in Fiji. Prior to joining USP, she was a science teacher educator at Lautoka Teachers College, the Fiji government's primary teacher training college at that time, and later at the Fiji National University. She has a special interest in the improvement of teaching and learning in science.

Ari Widodo Drs (IKIP Bandung), MEd (Deakin), Dr (Kiel) is a senior lecturer in the Faculty of Mathematics and Science Education at the Indonesia University of Education. His main research interests are analyses of science lessons, constructivism and conceptual change, and teacher professional development. He has published in academic journals and in books both in Indonesia and internationally. Currently, he is working on a research project on developing science teachers' pedagogical content knowledge (PCK) and how that knowledge impacts on lessons and students' learning.

Diana Rochintaniawati Dra (IKIP Bandung), MEd (La Trobe), Dr (Universitas Pendidikan Indonesia) is a senior lecturer and head of the International Program in Science Education, Faculty of Mathematics and Science Education, Indonesia University of Education. Her main research interests are science curriculum, resources for teaching science, and teacher professional development. She is presently working on a research project on the roles of computer animation on students' learning.

Riandi Drs (IKIP Bandung), MSc (Universitas Gajah Mada), Dr (Universitas Pendidikan Indonesia) is a senior lecturer in the Department of Biology Education, Faculty of Mathematics and Science Education, Indonesia University of Education. His main research interests are conceptual mapping, teaching media, and teacher professional development. He is currently working on a research project focused on developing science teacher pedagogical content knowledge (PCK) and how it affects lessons and students' learning.

Marie Perera BA (Peradeniya), MPhil, MSc (Colombo), PhD (Wollongong) is a professor of humanities education at the University of Colombo, Sri Lanka. She was Dean of the Faculty of Education until August 2013. Prior to that she was Head of the Department of Humanities Education and also served as Director of the Staff Development Centre of the University of Colombo. She is the principal author of the reports on the national assessment studies conducted by the National Education Research and Evaluation Centre of the University of Colombo. Marie was instrumental in obtaining a quality innovation grant for postgraduate studies for the Faculty of Education.

Saouma BouJaoude BS (American University of Beirut), MEd, EdD (Cincinnati) is a professor of science education at the American University of Beirut and is presently Director of the Center for Teaching and Learning. Saouma BouJaoude has published research in international journals and chapters in edited books in English and Arabic, is a member of several international science education research associations, and has served as the international coordinator and a member of the executive board of the National Association for Research in Science Teaching (NARST). He has also served as a consultant for educational institutions at the pre-university and higher education levels in Lebanon and the region.

Khalid Khan BSc (Sana'a), MEd (Al-Jazeera), PhD (Yemen UST) is an assistant professor in the Department of Education of the University of Hodiedah. He taught as a secondary teacher and then served as a physics inspector before going into teacher education at the Yemen University of Science and Technology and subsequently the University of Hodiedah. His interests revolve around quality assurance in higher education. He has published papers in this field as well as on teacher quality-related issues.

Muhammad Salahuddin BEd, MEd (Dhaka) is working as a researcher at the University of Dhaka's Institute of Education Research (IER) under the EIA-DU-OU (UK) Research Collaboration Programme. He has expertise in educational research, assessment, education policy, female education, and ethnic education.

Chan Oeurn Chey BSc (RUPP, Cambodia), MEng (KMUTT, Thailand), PhD (Linköping) graduated from the Department of Science and Technology, Institute of Technology, Linköping University, Sweden, January 2015. He works in several positions at the Royal University of Phnom Penh (RUPP): as teacher of undergraduate and graduate programs in the Faculty of Science, program coordinator for the Master of Science in physics, and trainer and team leader for the Cambodian team for the Asian Physics Olympiad (APhO), International Physics Olympiad (IPhO), and Astronomy Olympiad competitions. He also supervises many Cambodian Young Scientist candidates who compete in the Regional Congress on the Search for SEAMEO Young Scientists (SSYS).

Sitha Chhinh BEd (Phnom Penh), MEd, PhD (Hiroshima) graduated from the Graduate School for International Development and Cooperation (IDEC) of the University of Hiroshima in 2004 and is presently looking at development theory and practice from an educational perspective. His areas of teaching and research focus on policy and the political economy of education and how they contribute to the development of other social sectors. His current longitudinal research project involves an exploration of the political economy of schooling in Cambodia, with a particular focus on issues of quality and equity. He is a teacher and trainer in both qualitative and quantitative research and evaluation in education and development.

Charles Kivunja BA (Makerere), MSc (Nairobi), MAgrEcon (Sydney), MEd, PhD (Western Sydney) is a senior lecturer in pedagogy and leadership in the School of Education at the University of New England. He taught science in secondary schools in Uganda before moving to Australia, where he taught at several high schools and lectured at the Australian Catholic University in Sydney before taking up his current appointment.

John Sentongo BEdSc (Dar-es-Salaam), MSc (Makerere), PhD (University of the North) is a lecturer in the Department of Science, Technical, and Vocational Education at Makerere University. He has been involved in training secondary school science teachers at Makerere University for over 20 years. His major research interests include students' learning of science concepts and their alternative conceptions, gender and science education, and science teacher education.

Index